The Biology of Mangroves

Biology of Habitats
Series editors: M.J. Crawley, C. Little, T.R.E. Southwood, and S. Ulfstrand

The intention is to publish attractive texts giving an integrated overview of the design, physiology, ecology, and behaviour of the organisms in given habitats. Each book will provide information about the habitat and the types of organisms present, on practical aspects of working within the habitats and the sorts of studies which are possible, and include a discussion of biodiversity and conservation needs. The series is intended for naturalists, students studying biological or environmental sciences, those beginning independent research, and biologists embarking on research in a new habitat.

The Biology of Rocky Shores
Colin Little and J.A. Kitching

The Biology of Polar Habitats
G.E. Fogg

The Biology of Ponds and Lakes
Christer Brönmark and Lars-Anders Hasson

The Biology of Streams and Rivers
Paul S. Giller and Björn Malmqvist

The Biology of Mangroves

Peter J. Hogarth

OXFORD
UNIVERSITY PRESS

OXFORD
UNIVERSITY PRESS

Great Clarendon Street, Oxford OX2 6DP

Oxford University Press is a department of the University of Oxford.
It furthers the University's objective of excellence in research, scholarship,
and education by publishing worldwide in

Oxford New York

Athens Auckland Bangkok Bogotá Buenos Aires Calcutta
Cape Town Chennai Dar es Salaam Delhi Florence Hong Kong Istanbul
Karachi Kuala Lumpur Madrid Melbourne Mexico City Mumbai
Nairobi Paris São Paulo Singapore Taipei Tokyo Toronto Warsaw

and associated companies in Berlin Ibadan

Oxford is a registered trade mark of Oxford University Press
in the UK and certain other countries

Published in the United States
by Oxford University Press Inc., New York

British Library Cataloguing in Publication Data
Data available

Library of Congress Cataloging in Publication Data
Data available

ISBN 0 19 850223 0 (Hbk)
ISBN 0 19 850222 2 (Pbk)

1 3 5 7 9 10 8 6 4 2

Typeset in Baskerville
by J&L Composition Ltd, Filey, North Yorkshire
Printed in Great Britain
on acid-free paper by
Biddles Ltd,
Guildford and King's Lynn

Preface

To most people, mangroves call up a picture of a dank and fetid swamp, of strange-shaped trees growing in foul-smelling mud, inhabited mainly by mosquitoes and snakes. Mud, methane and mosquitoes are certainly features of mangrove forests—as, sometimes, are snakes. They are not sufficient to deter mangrove biologists from investigating an ecosystem of great richness and fascination.

Mangroves are an assortment of tropical and subtropical trees and shrubs which have adapted to the inhospitable zone between sea and land: the typical mangrove habitat is a muddy river estuary. Salt water makes it impossible for other terrestrial plants to thrive here, while the freshwater and the soft substrate are unsuitable for macroalgae, the dominant plants of hard-bottomed marine habitats. The mangrove trees themselves trap sediment brought in by river and tide, and help to consolidate the mud in which they grow. They provide a substrate on which oysters and barnacles can settle, a habitat for insects, and nesting sites for birds. Most of all, through photosynthesis, they supply an energy source for an entire ecosystem comprising many species of organism. Mangroves are among the most productive, and biologically diverse, ecosystems in the world.

In discussing this complex habitat, I start with the trees themselves, before considering the multiplicity of other organisms of the mangal. Some are invaders from the land, others from the sea: I discuss these components in turn, before attempting an integrated view, turning to more general questions of evolution, biodiversity (and whether it matters), and human impacts on mangrove habitats.

The productivity and diversity of mangroves are also of considerable economic value. The trees themselves are exploited for timber, firewood and grazing and a host of minor uses. Many of the animals found within the mangal are harvested for food: fish, crabs and prawns in particular. These uses are often crucial to local communities: many of the major mangrove areas are in countries with few other natural resources.

Finally, mangroves are currently threatened by many different human activities. Exploitation is often taken beyond the level of natural replacement. Inland irrigation schemes divert river water from coastal regions and mangroves suffer from the resulting increase in salinity. Pollution takes its toll. Deliberate clearance for the development of aquaculture or for the construction of airports takes place without consideration of what may be lost. And, of course, any significant rise in sea level resulting from global warming will mean encroachment on the relatively narrow zone within which mangroves flourish.

Destruction of mangrove habitats means a loss of the mangrove resource: as mangroves decline, so too do timber and charcoal production and fisheries, and the livelihoods of the people who depend on them. This seems almost too obvious to need pointing out. Cassandra was fated to predict the future and to have her predictions ignored; mangrove biologists sometimes feel they have a similar role.

The productivity and diversity of mangroves therefore make them of great interest to biologists and of considerable social and economic value, while degradation and destruction by human activities makes it more than ever essential to understand their significance. Mangrove research has advanced considerably in the last few years, and the time seems right for an attempt to present our current understanding of the ecology of mangrove habitats.

My aim in writing this book, then, is two-fold: to share my own enthusiasm for a remarkable ecosystem, and to explain how our understanding of that ecosystem is unfolding. Any author depends on the work of others, and I am grateful to numerous colleagues for their help in various ways. In particular, I should like to thank Larry Abele, Liz Ashton, Mike Gee, Rony Huys and Mohammed Tahir Qureshi. Colin Little and Cathy Kennedy of Oxford University Press made invaluable comments and suggestions: any errors that remain are entirely my own.

Writing books has its pleasures, particularly learning about areas of the subject with which one was previously not sufficiently familiar. It also has its disadvantages, and most authors would at some stage agree with the heartfelt—and, in this context, singularly appropriate—words of the great American naturalist John James Audubon: '*God . . . save you the trouble of ever publishing books on natural science . . . I would rather go without a shirt . . . through the whole of the Florida swamps in mosquito time than labor as I have . . . with the pen.*'[1] For sustaining me throughout the labours with the pen (and for joining me in the Malaysian swamps in mosquito time) I should especially like to express my gratitude to Sylvia, to whom this book is dedicated.

P.J.H.

York
March 1999

---

[1] Letter to J. Bachman, 1834, quoted by Alice Ford (1957): *The bird biographies of John James Audubon* (Macmillan, N.Y.) pp. vii–viii, and in Rosenzweig (1995).

Contents

Chapter 1   **Mangroves**   1
What is a mangrove?   1
Geographical distribution   2
Mangrove trees and their environment   3
    Adaptations to waterlogged soil   4
    Coping with salt   11
    The price of survival   16
    Inorganic nutrients   19
Reproductive adaptations   23
    Pollination   23
    Propagules   24
    Fecundity and parental investment   26
    Dispersal   27
    Why vivipary?   30
Why are mangroves tropical?   32

Chapter 2   **The mangrove ecosystem: can we see the wood for the trees?**   33
Form of the forest   33
    Riverine mangroves   33
    Tide-dominated mangroves   34
    Basin mangroves   34
    Special settings   35
Mangrove species zonation   36
    Why do mangroves show zonation?   37
How different are mangroves from other forests?   45
The mangrove environment: do mangroves make mud?   47

Chapter 3   **The mangrove community: terrestrial components**   51
Mangrove-associated plants   51
Animals from the land   52
    Insects   53
    Spiders   62
    Vertebrates   62

Chapter 4 **The mangrove community: marine components**   77
    Algae   77
    Fauna of mangrove roots   78
    Invertebrates   80
        Crustacea   81
        Mangrove molluscs   106
        Meiofauna   110
    Fish   112

Chapter 5 **Measuring and modelling mangroves**   116
    How to measure a tree   117
        Biomass   117
        Estimating production   120
    What happens to mangrove production?   123
        Microbial breakdown   125
        Crabs and snails   126
        Wood   127
        Role of sediment bacteria   128
        The fate of organic particles   129
        Predators   130
    Putting the model together   130

Chapter 6 **Comparisons and connections**   133
    How distinctive is the mangrove community?   133
    Exchange with the surroundings   134
        Outwelling   135
        The fate of mangrove exports   137
        Mangroves, seagrasses and coral reefs   140
        Commuters   142
        Larval dispersal and return   143
        Mangroves and fisheries   144
    Mangroves and salt marshes   149

Chapter 7 **Biodiversity and biogeography**   151
    What, if anything, is biodiversity?   151
    Regional diversity   152
    Origins   156
    Local diversity   160
    Genetic diversity   163
    Diversity of the mangrove fauna   165
    Diversity and ecosystem function   166

Chapter 8 **Impacts**   170
    Uses of mangroves   171

Direct uses   172
Indirect uses   173
Mangroves and coastal protection   174
Ecotourism   174
Sustainable management: the case of the Matang   175
Shrimps versus mangroves?   177
Mangroves and pollution   180
Hurricanes and typhoons   183
Mangrove rehabilitation   185
Mangroves and global climate change   185
Rise in atmospheric carbon dioxide   186
Global warming   187
Sea level rise   188
Mangroves of the Indus Delta: a case study   189
What are mangroves worth?   195
Have mangroves a future?   198

Further reading   200

Bibliography   201

Glossary   216

Index   219

1 Mangroves

Life is difficult between the tides. Marine organisms are exposed to the air, and to the risk of desiccation and over-heating. Terrestrial organisms face submersion, and fresh water ones are exposed to salt water. Organisms of the intertidal zone must cope with regular fluctuations in all of these aspects of their environment. The species that flourish in such an exacting environment are often highly specialised and do not occur elsewhere.

On tropical shores, these problems are made worse by high temperatures. At low tide, the linked dangers of over-heating and desiccation are greater, and through evaporation any water that remains may become even more highly saline than the sea. At high tide, the warmth of the water means lower oxygen levels in the water, and possible problems for gill-breathing animals.

Tropical rocky shores, as a result, are often virtually barren. In contrast, sandy or muddy shores are often rich in life. Most productive are the mangrove forests that fringe most tropical estuaries and inlets. Mangroves are trees and shrubs which flourish in the inhospitable habitat between land and sea. Not only do they flourish, they physically create a habitat for a rich community of other organisms, and through their photosynthesis provide the energy base of a remarkable ecosystem. The mangrove trees themselves, and the mangrove habitat, or mangal, are the subject of this book.

What is a mangrove?

Mangroves are defined as woody trees and shrubs which flourish in mangrove habitats (or mangals): almost, but not quite, a tautology. 'True', or 'exclusive' mangroves are those which occur only in such habitats, or only rarely elsewhere. There is, in addition, a loosely defined group of species often described as 'mangrove associates', or 'non-exclusive' mangrove species. These comprise a large number of species typically occurring on the landward margin of the mangal, and often in non-mangal habitats such as rain forest, salt marsh or lowland fresh water swamps. Many epiphytes also grow on mangrove trees: these include an assortment of creepers, orchids, ferns and other plants, many of which cannot tolerate salt and therefore grow only high in the mangrove canopy.

True mangroves comprise some 54 species in 20 genera, belonging to 16 families. Taxonomy is not now regarded as the most glamorous aspect of biology: nevertheless these figures indicate an important feature of mangroves as a group. They are taxonomically diverse. From this we can infer that the mangrove habit—the complex of physiological adaptations enabling survival and success—did not evolve just once and allow rapid diversification by a common ancestor. The mangrove habit probably evolved independently at least 16 times, in 16 separate families. The common features have evolved through convergence, not common descent.

The principal mangrove families and genera are listed in Table 1.1. Most families are represented by a small number of mangrove species, and they also contain non-mangrove species. However, of the 30-odd species that represent the major components of mangal communities, 25 belong to just two families, Avicenniaceae and Rhizophoraceae. These families dominate mangrove communities throughout the world.

Geographical distribution

Mangroves are almost exclusively tropical (Fig. 1.1). This suggests a limitation by temperature. Although they can survive air temperatures as low as 5°C, mangroves are intolerant of frost. Seedlings are particularly vulnerable. Mangrove distribution, however, correlates most closely with sea temperature. Mangroves rarely occur outside the range delimited by the winter position of the 20°C isotherm, and the number of species tends to decrease as this limit is

Table 1.1 The principal mangrove species, from Tomlinson (1986)

Family	Genus	Number of species
Major components		
Avicenniaceae	*Avicennia*	8
Combretaceae	*Laguncularia*	1
	Lumnitzera	2
Palmae	*Nypa*	1
Rhizophoraceae	*Bruguiera*	6
	Ceriops	2
	Kandelia	1
	Rhizophora	8
Sonneratiaceae	*Sonneratia*	5
Minor components		
Bombacaceae	*Camptostemon*	2
Euphorbiaceae	*Excoecaria*	2
Lythraceae	*Pemphis*	1
Meliaceae	*Xylocarpus*	2
Myrsinaceae	*Aegiceras*	2
Myrtaceae	*Osbornia*	1
Pellicieraceae	*Pelliciera*	1
Plumbaginaceae	*Aegialitis*	2
Pteridaceae	*Acrostichum*	3
Rubiaceae	*Scyphiphora*	1
Sterculiaceae	*Heritiera*	3

Fig. 1.1 World distribution of mangroves in relation to the January and July 20°C sea temperature isotherms. The broken lines indicate the biogeographical areas discussed in Chapter 7, and the arrows major ocean currents. The term Indo-Malesia describes a biogeographic region including India, southern China, Malaysia, Indonesia and other parts of S.E. Asia. (Reproduced from Duke, N.C. (1992). Mangrove floristics and biogeography. In *Tropical mangrove ecosystems*, ed. A.I. Robertson and D.M. Alongi, pp. 63–100. Copyright by the American Geophysical Union.)

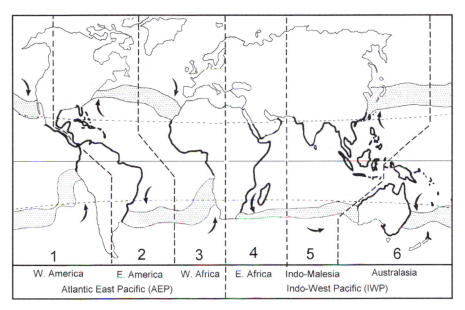

1	2	3	4	5	6
W. America	E. America	W. Africa	E. Africa	Indo-Malesia	Australasia
Atlantic East Pacific (AEP)			Indo-West Pacific (IWP)		

approached. In the southern hemisphere, ranges extend further south on the eastern margins of land masses than on the western, reflecting the pattern of warm and cold ocean currents. In South America, for example, the southern limit on the Atlantic coast is 33°S. On the Pacific coast, the cold Humboldt current restricts mangroves to 3°40'S.

In Australia and New Zealand, mangroves extend further south: the highest latitude at which mangroves are found is at Corner Inlet, Victoria, Australia, where a species of *Avicennia* occurs at 38°45'S. This extreme distribution may be due to local anomalies of current and temperature, or to the local evolution, for some reason, of an unusually cold-tolerant variety. Possible reasons for the limits in mangrove distribution are discussed on p. 32 and Chapter 7.

Mangrove trees and their environment

Typical mangrove habitats are periodically inundated by the tides (Fig. 1.2). Mangrove trees therefore grow in soil that is more or less permanently water-logged, and in water whose salinity fluctuates and may be as high as that of the open sea. How do they cope?

Fig. 1.2 Mangroves (*Avicennia marina*) at the edge of a tidal creek in the Indus delta, Pakistan.

Adaptations to waterlogged soil

The underground tissues of any plant require oxygen for respiration. In soils which are not waterlogged, gas diffusion between soil particles can supply this need. In a waterlogged soil, the spaces between soil particles are filled with water. Even when water is saturated with oxygen, its oxygen concentration is far below that of air, and the diffusion rate of oxygen through water is roughly 10 000 times less than through air (Ball 1988*a*). Waterlogged soils are therefore low in oxygen.

When oxygen movement into soil is severely limited, the oxygen that is present is soon depleted by the aerobic respiration of soil bacteria. Thereafter, anaerobic activity takes over. The result is that mangrove soils are often virtually anoxic.

A convenient method of measuring whether a soil is in an anaerobic state is the redox potential ('redox' being a telescoping of 'reduction' and 'oxidation'). This can be tested by insertion of a platinum electrode probe which senses the redox state of the surrounding soil. The redox scale is in millivolts (mV). A well oxygenated soil will have a redox potential above +300 millivolts (mV). As oxygen availability decreases, so does the redox potential, so that an anoxic mangrove soil may give a reading of −200 mV or lower.

Some unusual bacterially-mediated soil chemistry takes place at low soil oxygen levels. As the soil becomes progressively more anoxic and reducing, bacteria convert nitrate to gaseous nitrogen: this is typical of redox potentials of +200 to +300 mV. As oxygen declines further (redox potentials of +100 to +200 mV) iron is converted from its ferric (Fe^{3+}) to ferrous (Fe^{2+}) form. Since ferric salts are generally insoluble and ferrous ones soluble, this has the effect of releasing soluble iron and inorganic phosphates. These can be used by plants, although excessive uptake of iron is toxic. Finally, with redox potentials of -200 to -100 mV, sulphate is reduced to (toxic) sulphide, and carbon dioxide to methane (Boto 1984). The latter two reactions can result in mangrove mud being extremely pungent, as anyone who has worked in a mangrove swamp can testify, and certainly do not make the environment any more favourable for plant growth.

Mangrove trees have adapted to survive in such unpromising surroundings. The most striking adaptations are various forms of aerial root. The roots of most trees branch off from the trunk underground. In well oxygenated soil, there is little difficulty in obtaining the oxygen needed for respiration. This is not so in waterlogged soils, and special aerating devices are required. In growing *Rhizophora*, roots diverge from the tree as much as 2 m above ground, elongate at up to 9 mm d^{-1}, and penetrate the soil some distance away from the main stem (Figs 1.3, 1.4 and 5.1). As much as 24 per cent of the above-ground biomass of a tree may consist of aerial roots: the main trunk, as it reaches the ground, tapers into relative insignificance. Aerial roots sometimes branch, but apparently only if they are damaged, for instance by insect attack (Gill and Tomlinson 1975, 1977). Because of their appearance, and because they provide the main physical support of the trunk, the aerial roots of *Rhizophora* are often termed stilt roots. On reaching the soil surface, absorptive roots grow vertically downwards, and a secondary aerial root may loop off and penetrate the soil still further away from the main trunk. The aerial roots of neighbouring trees often cross, and the result may be an almost impenetrable tangle. This makes life very difficult for the mangrove researcher. Tomlinson (1986) quotes the world record for the 100 m dash through a mangrove swamp: 22 minutes 30 seconds.

The method of aerating the underground roots is understood best in the red mangrove *Rhizophora mangle* of Florida. Functionally, the aerial components can be divided structurally into more or less horizontal arches and vertical columns. These have no problems in achieving adequate gas exchange, at least at low tide. In contrast, the underground roots are in a permanently hypoxic, or even anoxic, environment. The columns have the role of supplying oxygen to the underground roots. Air passes into the column roots through numerous tiny pores, or lenticels, which are particularly abundant close to the point at which the column root enters the soil surface. It can then pass along roots through air spaces. Roots entering the soil are largely composed of aerenchyma tissue, honeycombed with air spaces which run longitudinally down the root axis (Fig. 1.5). These spaces are visible to the naked eye. Their extent and continuity can be demonstrated by blowing down a section of root as through a straw. It is quite easy to blow air through a section of mangrove root up to around 60 cm in length: powerful lungs are required for greater lengths.

The properties of the different components of mangrove roots are illustrated in Table 1.2. Roots in air, with no need to conduct gas internally, contain only a

Fig. 1.3 Different forms of mangrove root structure. (Reproduced with permission from Tomlinson, P.B. (1986). *The botany of mangroves*. Cambridge University Press.)

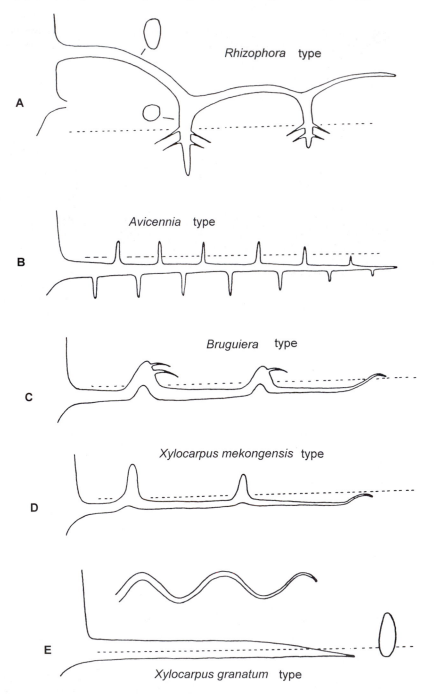

Fig. 1.4 Mangrove forest in western Peninsular Malaysia, showing the aerial roots of *Rhizophora*. There is little or no understorey vegetation, except in the sunlit clearing in the background, where it consists almost entirely of *Rhizophora* seedlings.

Fig. 1.5 Scanning electron micrograph of aerenchyma tissue near the tip of a pneumatophore of *Avicennia marina*. Magnifications: left ×50, right ×400. Micrographs by Fiona Graham, University of Natal. (Reproduced from Osborne, D.J. and Berjak, P. (1997). The making of mangroves: the remarkable pioneering role played by seeds of *Avicennia marina*. *Endeavour*, 21, 143–7, © 1997, with permission from Elsevier Science and the authors.)

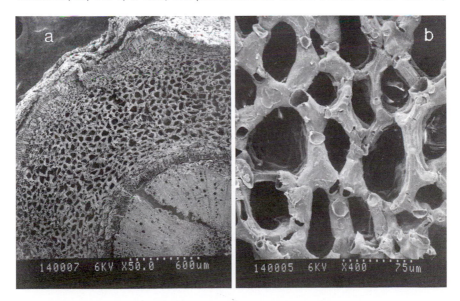

Table 1.2 Properties of *Rhizophora* roots in different environments, from Gill and Tomlinson (1977)

Environment	Lenticels	Chlorophyll	Gas space (% volume)
Air	+	+	0–6
Mud	–	–	42–51
Water (light)	–	+	22–29
Water (dark)	–	–	35–40
Drained sand	–	–	22–28

small volume of gas space, equivalent to less than 6 per cent of their volume. In anoxic mud, the gas space of a root may be more than half of the total root volume. Between these extremes, roots with different degrees of access to oxygen have intermediate proportions of gas space in their roots. In aerated soil such as drained sand, the need for internal gas movement is slight and gas space is relatively small. Waterlogged roots have a higher proportion of gas space, unless they are exposed to light. In this case, chlorophyll appears in the surface tissues and photosynthesis offsets the shortage of environmental oxygen, internal conduction is less important, and internal gas space is correspondingly less (Gill and Tomlinson 1977).

The importance of the lenticels for gas exchange has been demonstrated by measuring O_2 and CO_2 concentrations in the aerenchyma of *Rhizophora* roots. When the lenticels are occluded by smearing grease over the aerial portion of the root, O_2 declines continuously and CO_2 rises (Fig. 1.6). Control roots showed fluctuations related to tidal level (Scholander *et al.* 1955).

Looping aerial roots are typical of *Rhizophora*: other species have other forms of root architecture. In *Bruguiera* and *Xylocarpus* a shallow horizontal root periodically breaks the soil surface and submerges again, forming a knee root. In *Xylocarpus granatum* the upper surface of the horizontal root shows above the mud and grows in characteristic, sinuous curves (Fig. 1.3).

A different form of root architecture is shown by species such as *Avicennia*. Shallow, horizontal roots radiate outwards, often for a distance of many metres. At intervals of 15 to 30 cm vertical structures known as pneumatophores emerge and stand erect, up to 30 cm above the mud surface (Fig. 1.3). (In *Sonneratia*, pneumatophores may be a staggering 3 m in height.) A single *Avicennia* tree 2 to 3 m in height may have more than 10 000 pneumatophores (Fig. 1.7). The pneumatophores have abundant lenticels and, like the column roots of *Rhizophora*, extensive gas spaces to allow gas exchange with the underground tissues: aerenchyma may account for up to 70 per cent of root volume (Curran 1985). If the lenticels are sealed with Vaseline, oxygen concentration falls and eventually the underground roots asphyxiate.

Simple physical diffusion through the lenticels and along the aerenchyma is probably the main mode of gas movement in mangrove roots, but it may be supplemented by mass flow. One possibility is that changing water pressure during the rise and fall of the tides might alternately compress and expand

Fig. 1.6 Oxygen and carbon dioxide content of gas sampled from underground root of *Rhizophora*. When the lenticels are occluded with grease (lower curves) oxygen concentration declines and CO_2 rises compared with the controls (upper curves). Time and tidal state are also indicated. (Reproduced with permission from Scholander *et al.* (1955). Gas exchange in the roots of mangroves. *American Journal of* Botany, **42**, 92–8, copyright Botanical Society of America.)

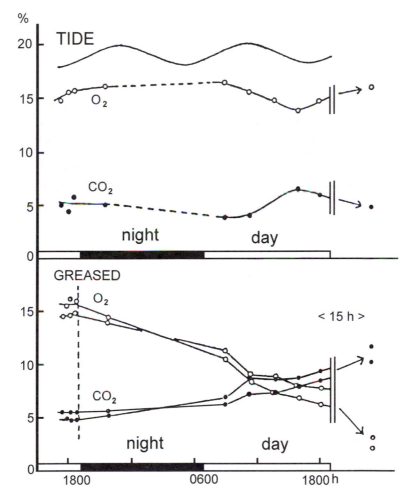

roots, resulting in successive inhalation and exhalation not unlike that of a vertebrate lung. However, this model predicts that pressure in the roots would increase at high tide and decrease at low. Direct measurements have shown that pressure changes in exactly the opposite way: increasing at low tide and decreasing at high.

There is a more convincing interpretation of the observed pressure changes which might provide a mechanism for the mass flow of air into a root to supplement diffusion. Lenticels are hydrophobic, so that while a root is covered

Fig. 1.7 Mangroves (*Avicennia marina*) at the edge of the Sinai Desert, Egypt, the most northerly limit of their distribution. The mass of pneumatophores in the foreground belong to the trees in the background, and are regularly covered by the sea.

by water they are in effect closed: neither air nor water can enter. Respiration removes oxygen from the air spaces and produces carbon dioxide. Because it is highly soluble in water, the carbon dioxide does not replace the volume of oxygen removed, and gas pressure within the root is therefore reduced. This is confirmed by direct measurement of gas composition in a submerged *Avicennia* root. After a root is covered by the tide oxygen within it falls, carbon dioxide levels do not increase to compensate, and pressure falls. When the tide recedes and the lenticels are again open, air is sucked in (Scholander *et al.* 1955).

Gas transport by mangrove root systems is so effective that it even aerates the surrounding soil. This has been shown by measuring the redox potential 15 cm below the surface of the soil both close to ($<$ 3 cm) and at some distance from (0.5 m) prop roots of *Rhizophora* and pneumatophores of *Avicennia* (Table 1.3). In this case *Rhizophora* clearly made the soil less hypoxic, although *Avicennia* had no significant effect (McKee 1993); in another, similar study, the opposite was found (Thibodeau and Nickerson 1986).

The relationships between structure and function of mangrove roots are relatively easy to study. Metabolic adaptation of roots to anoxia and waterlogging is less well understood. In aerobic conditions, most organisms carry out glycolysis, with the production of pyruvate. This is then disposed of via the Krebs (tricarboxylic acid) cycle and oxidative phosphorylation, with the production of ATP. In anaerobic conditions, this is not possible, and pyruvate is shunted sidewise to form lactate or ethanol. The former is favoured by animals: when aerobic

Table 1.3 Redox potentials (Eh: mV) and sulphide concentrations measured at depth of 15 cm close to (< 3 cm) or distant from (ca 0.5 m) *Rhizophora* prop roots or *Avicennia* pneumatophores, from McKee (1993)

	Redox potential (Eh: mV) ± s.e.		Sulphide (nM) ± s.e.	
	near	away	near	away
Rhizophora mangle (prop root)	−80 ± 21*	−203 ± 13	0.17 ± 0.08**	1.70 ± 0.34
Avicennia germinans (pneumatophore)	−154 ± 17	−200 ± 10	0.48 ±0.15**	1.83 ± 0.26

Significance: * $p < 0.05$, ** $p < 0.01$.

conditions return, the 'oxygen debt' can be repaid and normal oxidative pathways resumed. Ethanol is produced under anaerobic conditions by yeast, a phenomenon that has not gone unnoticed in human society at large.

In experimental conditions, hypoxic roots of *Avicennia* produce a three-fold increase in levels of the enzyme alcohol dehydrogenase (ADH), suggesting an adaptive switch to anaerobic metabolic pathways. There is no direct evidence of ethanol production. However, this may have been because the particular experimental method used to produce hypoxic conditions—a continuous flow of nitrogen over the roots—would have had the effect of effectively flushing any ethanol out of the experimental system (McKee and Mendelssohn 1987).

Root architecture differs according to species, but also to some extent depends on the prevailing conditions. In the northern Red Sea, for instance, small areas of *Avicennia* grow in sandy soil at the edge of the Sinai desert. Trees whose roots are regularly inundated by the tide have abundant pneumatophores. On the landward side a few rather stunted trees have contrived to establish themselves in apparently dry sand, where they seem to survive on underground seepage of fresh water. The surface soil is essentially dry sand through which air can readily penetrate. There is no need for special aerating structures: pneumatophores are entirely absent.

The rooting system of mangroves also has the function of anchorage. The nature of mangrove soil means that roots tend to remain close to the surface, and enter the seriously anoxic depths as little as possible. There are therefore no deeply anchored taproots, but the aerial roots of *Rhizophora* form a combination of guy-ropes and flying buttresses that provides a mechanically effective alternative. The aerial roots of *Rhizophora*, and the horizontally spreading system of *Avicennia*, provide effective anchorage in a soil that is often fluid and unstable, as well as being anoxic. Compared with most tree species, the roots of mangroves comprise a relatively high proportion of the tree, which will enhance anchorage.

Coping with salt

Mangroves typically grow in an environment whose salinity is between that of fresh water and sea water, sea water comprising approximately 35 g l^{-1} salt, (including 483 mM Na^+ and 558 mM Cl^-). This means an osmotic potential of -2.5 MPa, and water must be taken in against this pressure. In some circumstances, mangroves even find themselves in hypersaline conditions, and the problem of water acquisition is correspondingly worse: in the Indus delta of Pakistan, for instance, evaporation raises the prevailing salinity to twice that of the sea. The problem may be compounded by fluctuations in salinity caused by the tide. Variation in salinity may be more difficult to cope with than high salinity itself.

Mangroves deploy a variety of means to cope with this unpromising environment. The principal mechanisms are exclusion of salt by the roots, tolerance of high tissue salt concentrations and elimination of excess salt by secretion. The interplay between these is complex, and not clearly understood. The requisite field experiments and measurements are not easy to carry out, and laboratory experiments tend to involve isolating one factor and ignoring interactions. Moreover, mangroves in natural situations display an ingenuity that makes interpretation very difficult. Trees which appear to be surrounded by saline, or even hypersaline, water may be satisfying their requirements from fresh water seepage, or from a subterranean lens of brackish water.

It may seem impossible to tell which of its alternative sources of available water a mangrove tree is actually using. An ingenious series of experiments has exploited the observation that ocean water and fresh water differ in the ratio of two isotopes of oxygen, ^{18}O and the more abundant ^{16}O. The isotope ratio in plant xylem reflects that of the water taken up by the roots. In studies of coastal vegetation in southern Florida, mangroves took water from various sources: in some cases trees whose surface roots were bathed in sea water were relying for their water uptake on brackish or fresh water in the surface layer of the soil. *Rhizophora* seems to depend entirely on water in the top 50 cm of soil, in which 70 per cent of its fine roots are deployed (Lin and Sternberg 1994; Sternberg and Swart 1987).

In *Aegiceras* and *Avicennia*, 90 per cent of salt is excluded at the root surface, rising to 97 per cent as the salinity of the environment increases. The concentration of the xylem sap is about one tenth of that of sea water (Tomlinson 1986). Although the exclusion mechanism is not understood, it appears to be a simple physical one. Negative hydrostatic pressure is generated within the plant, largely by transpiration processes. This is sufficient to overcome the negative osmotic pressure in the environment of the roots. Water is therefore drawn in. Unwanted ions—and other substances such as dye molecules—are excluded (Moon *et al.* 1986).

The physical nature of this desalination process was demonstrated by Scholander (1955). Decapitated *Avicennia* and *Ceriops* seedlings were positioned in a 'pressure bomb', a container in which the roots, immersed in sea water, could be subjected to pressure by introducing compressed nitrogen while the cut stem was exposed to the atmosphere. With a positive pressure on the roots of 4 to

4.5 MPa (40–45 atmospheres), the cut stems exuded about 4 ml h^{-1} of water whose salinity was 0.2 per cent NaCl—around 5 per cent of that of the water surrounding the roots at the start of the experiment. As the water exuded over a number of hours was greater in volume than the total water capacity of the seedling, the desalinated water must have come largely from the environment and not from within the seedling itself. In the water surrounding the roots, the salt concentration increased ten-fold, showing that salt was being excluded rather than accumulated in some part of the seedling.

When seedlings were poisoned with either carbon monoxide or the metabolic inhibitor dinitrophenol, the rate of desalination was the same, indicating that the ultrafiltration is a physical process that does not depend on metabolic processes.

Even with exclusion of most of the salt, concentration of sodium and chloride ions within the plant tissues is higher than in non-mangrove species. High salt concentrations are known to inhibit many enzymes. Intracellular mangrove enzymes are protected partly by being more resistant to this inhibition, and partly by partitioning of solutes within different cellular components. Sodium and chloride ions are at high concentration within cell vacuoles, but are largely excluded from the cytoplasm itself. To avoid the consequences of an osmotic imbalance within the cell, high cation concentrations in the vacuole are balanced by high concentrations of non-ionic solutes in the cytoplasm. Various substances probably contribute, particularly glycinebetaine, proline and mannitol (Wyn Jones and Storey 1981).

In several mangrove species, among them *Avicennia*, *Rhizophora*, *Sonneratia* and *Xylocarpus*, sodium chloride is also deposited in the bark of stems and roots. Several species deposit salt in senescent leaves, which are then shed. This may help to remove salt from metabolic tissues. The deciduous mangroves *Xylocarpus* and *Excoecaria* appear to dump excess salt in this way in preparation for a new growing and fruiting season (Hutchings and Saenger 1987).

A number of species of mangrove possess salt glands on their leaves. Leaves of *Aegiceras* and *Avicennia* taste strongly salty when licked, and often carry clearly visible deposits of salt crystals. The lower leaf surface of *Avicennia* is densely covered with hairs, which raise the secreted droplets of salty water away from the leaf surface, preventing the osmotic withdrawal of water from the leaf tissues (Osborne and Berjak 1997).

The glands of *Avicennia* appear to be formed only in response to saline conditions, while in *Aegiceras*, and most other gland-bearing species, they are present regardless of the environmental salinity. Salt glands resemble each other closely across a range of species, presumably as the result of convergent evolution. They also closely resemble glands that secrete other materials, such as nectar, suggesting a possible evolutionary origin. Although the structure of the glands is well known, the secretion mechanism is still not well understood. The secretory cells are packed with mitochondria, suggesting intense metabolic activity. Secretion of salt can be prevented by metabolic inhibitors: it requires energy (Hutchings and Saenger 1987; Tomlinson 1986).

Exclusion, tolerance and secretion are used with different emphasis by different species, and within a species under different environmental conditions. Table 1.4 summarises the known occurrence of these mechanisms in a range of mangrove species.

Given the range of methods of coping with salt, it is not surprising that mangrove species differ in the extent of their salt tolerance. Some species, such as *Sonneratia lanceolata*, show maximal growth in solutions between fresh and 5 per cent sea water, while most species seem to grow best at 5 to 50 per cent sea water (Fig. 1.8). A few species (*Ceriops decandra, Sonneratia alba*) grow very poorly in fresh water, and *Bruguiera parviflora* and *Ceriops tagal* propagules fail to grow at all. The addition of as little as 5 per cent sea water to the culture solution promotes vigorous growth in these species (Ball 1988a; Hutchings and Saenger 1987).

The way in which salinity tolerances and preferences are tested can influence the conclusion. Culture experiments which simply vary salinity levels ignore interactions between salt tolerance and other environmental variables. For example, salinity interacts with the nature of the substrate: seedlings of *Avicennia* transplanted into sand and exposed to hypersaline conditions (three times the concentration of sea water) for 48 h shed all of their leaves. When the sand was mixed with 5 to 10 per cent clay, the result was some curling and discoloration of leaves: with a higher proportion of clay in the soil the seedlings appeared to survive unscathed. The mechanism of this interaction is not clear, but it is probably connected with ion adsorption on to the clay particles and a consequent reduction in the effective salinity of the soil water (McMillan 1975).

Salt tolerance is also likely to show the same degree of intraspecific variation as any other physiological trait. Individuals collected from different environments, or from different parts of a species' geographical range, may respond differently

Table 1.4 Mechanisms of coping with salt and their known distribution in a range of mangrove species (various sources)

Species	Exclude	Secrete	Accumulate
Acanthus		+	
Aegialitis	+	+	
Aegiceras	+	+	
Avicennia	+	+	+
Bruguiera	+		
Ceriops	+		
Excoecaria	+		
Laguncularia		+	
Osbornia	+		+
Rhizophora	+		+
Sonneratia	+	+	+
Xylocarpus			+

Fig. 1.8 Growth of *Avicennia* seedlings at various concentrations of sea water, measured as dry weight per plant. (Reproduced with permission from Hutchings, P. and Saenger, P. (1987). *Ecology of mangroves.* University of Queensland Press.)

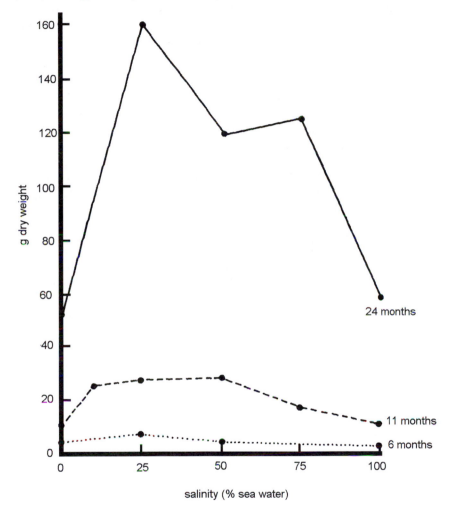

to varying salt concentrations. Developmental changes also occur: newly released *Avicennia* propagules generally grow best in 50 per cent sea water, but at later stages the optimal salinity is lower (Ball 1988*a*).

It is factors like these that probably explain some of the apparent discrepancies in published accounts: for example the optimal salinity for growth of *Avicennia marina* has been quoted as 10 to 50 per cent, 20 per cent, 25 per cent or 50 per cent sea water, and for *Rhizophora mangle* has been given as 25 per cent and 100 per cent by different authors (data in Field 1984).

If any general conclusion can be drawn, it is probably that mangrove species vary more in the range of their tolerance than in the salinity for optimal growth.

The price of survival

Salt tolerance is costly. Mangrove roots take up water much less freely than those of other plants. Consequently, a greater relative root mass is needed to satisfy the demand for water. In *Avicennia* and *Aegiceras* the root: shoot ratio is relatively high, and increases with increasing environmental salinity (Fig. 1.9) (Ball 1988*b*; Saintilan 1997). Materials committed to root growth cannot be invested in alternative structures, such as leaves.

Even in conventional plants, ion uptake accounts for an appreciable proportion of the energy costs of the plant. In maize, for instance, about half of the root respiration is probably due to ion uptake, and about 20 per cent of the total plant respiration is attributable to the costs of ion uptake and transport. These costs are, if anything, likely to be higher in mangroves. The leaves of *Avicennia* seedlings respire more rapidly in sea water than in fresh water or 25 per cent sea water. Respiration in the roots falls by 45 per cent, possibly due to limited supply of photosynthetic products in these conditions (Ball 1988*b*; Field 1984).

The objective of a plant is to assimilate carbon in photosynthesis. This cannot be done without the expenditure of water. In a mangrove forest, water is not in short supply but its acquisition is costly because of the problems of coping with salt. Although mangrove trees are surrounded by water, they cannot afford to be extravagant in its use. A useful index of water economy is the water use

Fig. 1.9 Variation in ratio of below ground to above ground biomass with increasing salinity for communities of *Avicennia* (squares) and *Aegiceras* (circles). Salinity is in parts per thousand (ppt). (Reproduced with permission from Saintilan, N. (1997). Above- and below-ground biomasses of two species of mangrove on the Hawkesbury River, New South Wales. *Marine and Freshwater Research*, 48, 147–52, © CSIRO.)

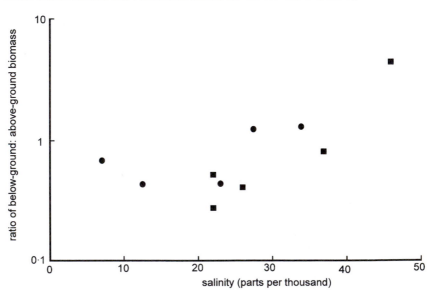

efficiency, the ratio of carbon assimilated to water used. In mangroves this is higher than in comparable non-mangrove trees, particularly in the more salt-tolerant species.

Carbon dioxide enters, and water is transpired, through the leaf stomata. At high soil salinities, stomatal conductance—the passage of gases through the stomata—is reduced. This conserves water by reducing transpiration, but also reduces CO_2 uptake (and growth). The balance between the two is such that, except under extreme conditions, just sufficient water is expended to maintain the carbon assimilation rate very near the photosynthetic capacity of the leaf.

Parsimonious use of water leads to other problems. Photosynthesis proceeds most rapidly in *Rhizophora* at a temperature of 25°C, falling off sharply above 35°C. The optimal temperature is typical of the air temperature within a mangrove forest. However, to maximise photosynthesis a leaf must position itself broadside-on to the sun. Maximising incident light, unfortunately, also maximises heat gain, and the temperature of a leaf in this position rapidly rises to 10–11°C above air temperature. One way of reducing leaf temperature would be to increase the transpiration rate and lose heat by evaporation. Mangroves cannot afford to do this. Instead, they tend to hold their leaves at an angle to the horizontal, so minimising heat gain. The angle varies from about 75° in leaves with greatest exposure to the sun, to 0° (horizontal) in leaves in full shade. Cooling is also enhanced by leaf design. Small leaves lose more heat by convection than large ones: leaves exposed to full sunlight, and heat-stressed, are smaller than those that are shaded. Leaves also tend to be smaller in the more salt-tolerant species, where water economy must be more stringent (Ball 1988*a*; Ball *et al.* 1988). Such constraints on leaf morphology may explain the convergent similarity between the leaves of different mangrove species.

Mangroves must therefore achieve a balance between minimising water expenditure, holding down leaf temperature and maximising carbon dioxide acquisition and growth. The trade-off between growth and salt tolerance is neatly shown by comparing two species of *Sonneratia*, *S. lanceolata* and *S. alba*. The former grows in salinities of up to 50 per cent that of sea water, while the latter has a broader tolerance and can grow in 100 per cent sea water. Both species show optimal growth at 5 per cent sea water. However, at optimal salinity the growth of *S. alba*, measured as biomass, height or leaf area, is less than half that of the less salt-tolerant species. A species can apparently opt for salt-tolerance or for rapid growth, not both. An ecological implication of this comparison, of course, is that *S. lanceolata* will be the successful competitor even at a salinity that is optimal for both species, but will not be able to grow or compete at high salinities. This is consistent with the actual distribution of the species along natural salinity gradients (Ball and Pidsley 1995).

The elaborate aerial root structures that enable mangroves to cope with anoxic soils (p. 4) represent constructional and maintenance costs to the plant.

Increasing anoxia increases this investment: as already noted, pneumatophores increase in number with decreasing soil oxygenation.

Mangroves therefore cope with the environmental stresses of salt and water-logging, but at the expense of growth, leaf area and photosynthesis. In extreme conditions, growth may be so restricted that dwarfing occurs. Large areas of the Indus delta in Pakistan, for instance, are covered with dwarf *Avicennia* less than 0.5 m in height: not seedlings, but adult trees possibly decades old (p. 195).

Given the costs of tolerance of high salt and low oxygen levels, it is not surprising to find that mangroves tend to avoid extremes of both simultaneously. Fig. 1.10 shows how a number of Australian mangrove species are distributed in relation to different levels of salinity and waterlogging. Species found at parti-cularly high salinities do not occur at high levels of waterlogging, and *vice versa*. These limits to distribution are narrower than the extremes that the species could actually survive, so the actual distributions reflect interspecific competi-tion as well as physiological tolerance.

Fig. 1.10 Tolerance range of salinity (parts per thousand, ppt) and per cent of time inundated for 17 Australian mangrove species. (From data in Hutchings, P. and Saenger, P. (1987). *Ecology of mangroves.* University of Queensland Press.)

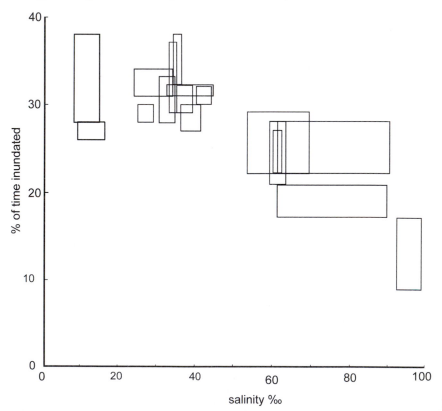

Inorganic nutrients

In addition to water, plants also require an adequate supply of mineral nutrients. The most important of these are nitrogen and phosphorus, in the form of inorganic nitrate and phosphate. Possible sources of inorganic nutrients are rainfall, fresh water from rivers or as runoff from the land, tide-borne soluble or particle-bound nutrients, the bacterial fixation of atmospheric nitrogen and release by microbial decomposition of organic material. Overall levels of these elements are affected by a complex and shifting balance between different sources: their actual availability to the mangroves depends crucially on the nature of the mangrove soil, and on microbial activity within it. Mangrove trees and their soil have a complicated relationship.

The nutrient contribution made by rainfall is probably small: mangrove areas with heavy rainfall (and some of those with very little rain) tend to have large fresh water inflow from rivers or land runoff. Potentially, mangroves also have a nutrient source in regular tidal inundations. The relative importance of these sources is not clear, and will undoubtedly vary greatly between different mangrove areas; and, of course, nutrients may be lost as well as gained to river water and the tides. The few cases that have been analysed in any detail indicate that most nutrients available to mangroves are of terrestrial origin (Lugo *et al.* 1976), and that mangrove habitats are more likely to have a net gain of nutrients than a net loss (e.g. Rivera Monroy *et al.* 1995).

Nitrogen and nitrogen fixers

Fixation of atmospheric nitrogen (N_2) by root-associated bacteria is an important source of nitrogen for many terrestrial plants. The situation with mangroves is not clear. One method of assessing the nitrogen-fixing ability of soil or root material depends on measuring levels of nitrogenase enzymes, responsible for reducing gaseous N_2 to ammonia (NH_3). Nitrogenases also reduce acetylene to ethylene, hence acetylene reduction can be taken as a measure of nitrogenase activity, and of nitrogen-fixing capacity.

On this basis, it appears that some nitrogen-fixing ability is associated with the roots of *Rhizophora mangle*, *Avicennia germinans* and *Laguncularia racemosa* from Florida, probably associated with the bacterium *Desulfovibrio* (Zuberer and Silver 1975). Other studies have given more equivocal results, and indeed the applicability of the acetylene reduction method has been questioned (Boto 1982).

Some species of mangrove (*Laguncularia*, *Lumnitzera* and *Aegiceras* among them) form leaf nodules containing bacteria. These may well be capable of nitrogen fixation, as in other angiosperm species, in which case the association represents a further potential source of nutrient to the tree (Hutchings and Saenger 1987).

An intriguing association of a different kind has recently been shown between Caribbean *Rhizophora mangle* and the sponges *Tedania* and *Haliclona*, which grow on its roots. Epiphytes and epizoites generally have adverse effects on the mangroves on which they grow, because they block lenticels and impede gas exchange. In this case, it appears that both partners benefit: soluble nitrogenous compounds pass from the sponge to the plant, and carbon compounds in the reverse direction. Sponges grow up to 10 times faster on mangrove roots than on inert substrates, and mangrove roots respond to the presence of sponges with rapid proliferation of rootlets ramifying through the sponge tissue (Ellison *et al.* 1996).

Although there are a few reports of mycorrhizal fungi associated with mangroves, it does not seem that mycorrhizal associations play a major role in mangrove ecosystems.

The contribution made by nitrogen-fixing bacteria, and by other organisms such as sponges, is not great; but it may be significant that inter-species relationships seem to revolve round nitrogen availability. This suggests that mangroves are often nitrogen-limited and any additional sources of nitrogen, even though quantitatively small, may be valuable.

The availability of inorganic nitrogen depends also on a complex pattern of activity of bacteria within the soil. Mangrove soil is largely anoxic, apart from a very thin aerobic zone at the surface. Ammonia is produced—either by nitrogen-fixation or decomposition of organic matter—principally in the anoxic zone. It diffuses upwards into the aerobic zone. Although some may be lost to the atmosphere, the bulk is probably oxidised by aerobic bacteria, first into nitrite, then into nitrate ions. This process is termed nitrification. Nitrate then diffuses back down through the anoxic layer. Here its fate depends on circumstances. Nitrate may be taken up by mangrove roots; it may be assimilated by bacteria and immobilised; or it may be reduced by further (anaerobic) bacterial action into either gaseous nitrogen or nitrous oxide (N_2O). In this last case, the gas is likely to diffuse through the soil and may be lost to the atmosphere (Boto 1982; Hutchings and Saenger 1987). These processes are summarised in Fig. 1.11.

The amount of nitrate actually available to mangrove roots depends on the balance between these processes, hence on the degree of oxygenation of the soil. This, in turn, depends on the inundation regime and on soil composition

Fig. 1.11 Chemistry of nitrogen compounds in mangrove soil. Solid arrows: conversion, broken arrows: diffusion. (Reproduced with permission from Boto, K.G. (1984). Waterlogged soils. In *The mangrove ecosystem: research methods*, Snedaker, S.C. and Snedaker, J.G. (eds). pp. 114–130, p. 117, Fig. 7.2 © UNESCO 1984.)

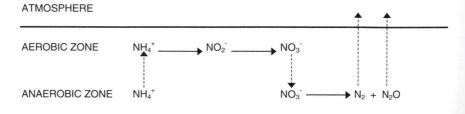

ATMOSPHERE

AEROBIC ZONE NH_4^+ ⟶ NO_2^- ⟶ NO_3^-

ANAEROBIC ZONE NH_4^+ NO_3^- ⟶ $N_2 + N_2O$

(p. 4). Soil oxygenation is also affected by animal burrows (p. 106) and by gas leakage from mangrove roots (p. 10), while tannins and other substances exuding from mangrove roots or leaching out of decomposing leaves may inhibit nitrification (Boto 1982).

Phosphorus

The position of phosphorus is just as complex. In Australian estuaries, considerable phosphate is removed from the water column and adsorbed on sediments as insoluble ferric phosphate. In anaerobic conditions, this is reduced to ferrous phosphate. These processes are impeded by poisoning mangrove sediments with formalin, and so must result from metabolic activity of bacteria, and not from physical or chemical processes (Hutchings and Saenger 1987). Ferrous phosphate is soluble, so phosphate may be leached out of the soil and lost. This depends on soil porosity: in fine clayey soils interchange between pore water and the water column is less, and such soils may therefore be richer in phosphate and support more flourishing mangroves.

Do nutrients limit growth?

Measurements of actual pore water nitrate and phosphate levels in mangrove soils show a great deal of variation, but in general levels are low (Alongi 1996; Alongi *et al.* 1992). The best evidence, however, that mangroves are nutrient limited comes from observing the response to raised nutrient levels. A comparison of two *Rhizophora mangle* islands in Florida showed greater mangrove growth on the one with a breeding colony of more than 100 egrets and pelicans, presumably due to the resulting nutrient input in the form of guano (Onuf *et al.* 1977).

Mangrove productivity has been shown to correlate with soil phosphate levels on a mangrove site in northern Australia (Fig. 1.12). In this case, however, the results do not unambiguously prove that nutrient limitation restricts growth. Both tree productivity and phosphate levels correlate also with topographic height of the sample sites. Also, phosphate exists in different forms, and it is not clear that the method of chemical analysis used to measure soil phosphate (or, indeed, any method of analysis) necessarily reveals the level of phosphate actually available to the trees (Boto 1984).

A more direct approach to testing whether mangroves are nutrient limited would be to fertilise plots artificially and compare tree growth or productivity with appropriate control plots. This has been done, also at a site in northern Australia. Three sites were chosen along a transect across Hinchinbrook Island, Queensland. Two were near the ends of the transect, close to the opposite edges of the island; the third was in the middle of the island, at a more elevated position. One edge site (Edge 1) and the middle site were then fertilised with phosphate, the remaining edge site (Edge 2) with ammonium. Each site was paired with an untreated control plot, and at all plots, treated and control,

Fig. 1.12 Variation of mangrove productivity and soil phosphate with topographic height on an island in northern Australia. (Reproduced with permission from Boto, K.G. Waterlogged soils (1984). In *The mangrove ecosystem: research methods*, (ed. Snedaker, S.C. and Snedaker, J.G.). pp. 114–130, p. 128, Fig. 7.5 © UNESCO 1984.)

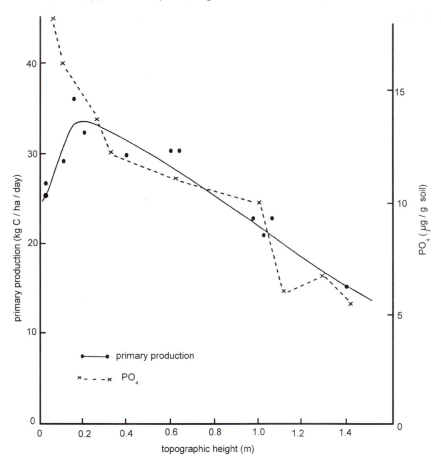

mangrove productivity was measured by a pair of litter collectors, for a full year before any treatment, and a full year following the start of treatment. New leaves form as a bud enclosed in a pair of stipules, which are shed as the bud expands. Of the various components of the litter, such as leaves, twigs and reproductive products, stipule fall correlates most closely with the rate of appearance of new leaves, and was therefore taken as an indication of productivity. The dominant mangroves at all sites were species of *Rhizophora*.

The key results of this study are shown in Table 1.5. The only significant changes are increases in productivity at the middle phosphate-treated site, and the edge site treated with ammonium. The implication is that the mangroves were nitrogen limited at one low-lying site (Edge 2), and phosphorus-limited at the higher (middle) site, but not at the other low-lying site (Edge 1). Other studies in Florida

Table 1.5 Leaf production following different nutrient treatments, from Boto and Wellington (1983)

Site	Treatment	Stipule fall before treatment g m^{-2} d^{-1}	Stipule fall after treatment g m^{-2} d^{-1}	% difference	Significance
Edge 1	phosphate	0.272	0.284	+4.4	n.s.
Control	none	0.409	0.391	−4.4	n.s.
Middle	phosphate	0.327	0.408	+24.8	*
Control	none	0.355	0.408	+14.9	n.s.
Edge 2	ammonium	0.402	0.490	+21.9	*
Control	none	0.498	0.452	−9.1	n.s.

* = $p < 0.05$.

and the Caribbean detected phosphate-limited growth of mangroves, but found little evidence of nitrogen limitation (Feller 1995, 1996).

Why the difference between high and low levels at Hinchinbrook Island? More sediment is deposited at lower shore levels, and sediment is probably a major source of phosphate (Boto and Wellington 1983). Lower shores are also inundated more frequently, and for longer, than higher shores. As a result, they are more waterlogged and anoxic; and at more negative redox potentials denitrification will convert usable nitrate into gaseous N_2 and N_2O, rapidly lost to the atmosphere (Fig. 1.11, p. 20).

The mangrove environment, then, is in general poor in nutrients, to the extent that growth of the trees may be restricted. Acquisition of sufficient nitrogen and phosphorus (and, presumably, of other essential substances as well) depends on complex interactions between the soil environment, microbial activity within it and on its aerobic state.

Reproductive adaptations

Pollination

The mode of pollination is generally inferred from observation of flower and pollen morphology, and of visits to flowers by possible pollen vectors. Species which produce large amounts of light powdery pollen, which do not smell attractive and which do not produce nectar, are probably wind-pollinated. In contrast, flowers producing a great deal of nectar to attract animals are likely to be vector-pollinated.

On this basis, *Rhizophora* appears to be wind-pollinated, although bees do visit the flowers regularly, and some nectar may be produced to encourage them. *Sonneratia* is served by bats or hawk moths (see p. 75, Chapter 3), large-flowered species of *Bruguiera* by birds and small-flowered species by butterflies. Bees of various types are probably the main pollinators of *Acanthus*, *Aegiceras*, *Avicennia*,

Excoecaria and *Xylocarpus. Nypa* may be pollinated either by small bees, or by flies of the family Drosophilididae, related to the familiar fruit fly (Hamilton and Murphy 1988). The flowers of these species are typically clustered, attractively scented and produce small quantities of nectar, presumably insufficient to attract birds or bats. *Ceriops* and *Kandelia*, and probably other species, are pollinated by a motley collection of small insects (Tomlinson 1994).

Propagules

All mangroves disperse their offspring by water. A distinctive feature of the majority of mangrove species is that they produce unusually large propagating structures, or propagules. This term is used because in most mangrove species what leaves the parent tree is a seedling, not a seed or a fruit. After pollination the growing embryo remains on the parent tree, and is dependent on it, for a period that often stretches to many months. The phenomenon is known as vivipary. (For some reason the equivalent in animals is termed 'viviparity'.)

Reproduction in orthodox species of plant involves dehydration of the developing seed, accompanied by cessation of DNA synthesis and cell division. These processes seem to be induced by the hormone abscisic acid (ABA). After it is shed, the seed can tolerate desiccation; the process has evolved in a varying environment to allow the embryo to wait, in a quiescent state, for favourable conditions to recur. When conditions are again suitable, the seed germinates, one of the first signs being the outgrowth of roots.

In mangrove species which show vivipary, little ABA is present, the developing propagule does not dehydrate, and DNA synthesis and cell division continue, producing an extended embryonic axis (the hypocotyl) or enlarged cotyledons. Root growth is suppressed as long as the propagule remains attached to its parent. Mangroves differ in this respect from the majority of plants, where seeds are produced which are resistant to desiccation, and which enter a quiescent stage. When the mangrove propagule parts from its parent, there is no period of quiescence, and no ability to resist drying out. Mangrove propagules are poised for germination and root outgrowth whenever the opportunity offers. These features appear to be common to viviparous mangrove species, and represent evolutionarily convergent adaptations to the mangrove environment (Farnsworth and Farrant 1998; Osborne and Berjak 1997). Conditions may be harsh, but they do not show much seasonal variation. In particular, the environment is never short of water, so the strategy of producing quiescent seeds would not be advantageous.

The most advanced form of vivipary is shown by *Rhizophora* and other members of the family Rhizophoraceae. After fertilisation the embryo develops within a small fruit. As the embryonic axis, or hypocotyl, elongates, it bursts through the surrounding pericarp and develops into a spindle-shaped structure which soon dwarfs the fruit (Fig. 1.13). Unlike conventional seeds, there is no period of dormancy, and growth is continuous (Sussex 1975; Tomlinson 1986).

Fig. 1.13 Stages in the development of a *Rhizophora* propagule. 1: branch with mature propagules, one of which is detached, 2: details of relation between cotyledonary collar and plumule of seedling, 3–6 stages in elongation of the hypocotyl. The swollen region, proximal to the hypocotyl, corresponds to the fruit, and the hypocotyl to the seedling. (From Gill, A.M. and Tomlinson, P.B. (1969). Studies on the growth of red mangrove (*Rhizophora mangle* L.). 1. Habit and general morphology. *Biotropica*, 1,1–9, copyrighted 1995 by the Association for Tropical Biology, P.O. Box 1897, Lawrence, KS 66044–8897. Reprinted by permission.)

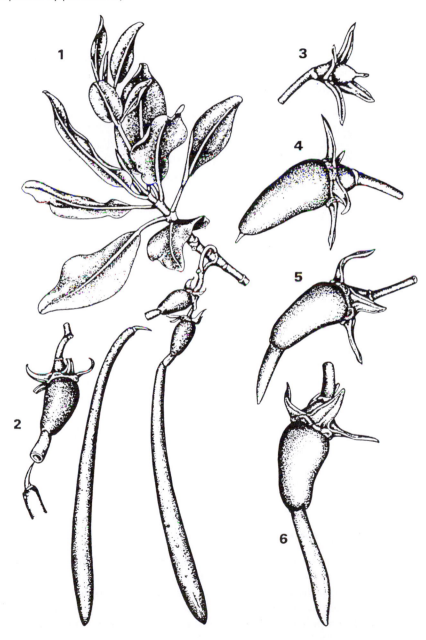

While still attached to the parent, the developing seedling develops chlorophyll and actively photosynthesises. The water and necessary nutrients are, of course, supplied by the parent. Carbohydrates are also translocated from the parent, and this is probably more important in the later stages of seedling development, as the rate of starch accumulation increases at a time when seedling photosynthesis is actually decreasing (Gunasekar 1993; Pannier and Fraíno de Pannier 1975).

Salt concentrations decline between the pedicel and the cotyledons, and then again between cotyledons and hypocotyl, and decline still further towards the tip of the hypocotyl, reaching levels of less than a third of those found in the pedicel (Lötschert and Liemann 1967). Tissues of the seedling are thus preserved from premature exposure to high salt levels.

The protruding hypocotyl may reach a considerable length. In the New World red mangrove *Rhizophora mangle* it reaches 20 to 25 cm and may weigh about 15 g; on *R. mucronata* of South-east Asia attached seedlings of 1 m have been recorded (Sussex 1975; Tomlinson 1986). Eventually the hypocotyl detaches from the residual fruit and, leaving its cotyledons behind, falls from the parent tree to start its independent life.

Aegiceras, *Avicennia* and a number of other mangrove species show a broadly similar form of reproduction, known as cryptovivipary, in which germination and embryonic development take place on the parent tree: in this case, however, the developing hypocotyl fails to penetrate the pericarp. Otherwise, the relationship between parent and offspring is much the same as in *Rhizophora*. The seedling again receives photosynthetic products from its parent, and salt concentrations in the hypocotyl are appreciably lower than those in the adult, or in the intervening tissues of the fruit (Bhosale and Shinde 1983; Joshi *et al.* 1972).

The final group of mangroves reproduce and disperse by more or less conventional seeds, ranging in size from a few millimetres in length to the massive fruit of the aptly named cannonball tree *Xylocarpus*, which weighs up to 3 kg and contains up to 20 individual seeds. The different forms of mangrove reproduction, and propagule sizes, are shown in Fig. 1.14 (Tomlinson 1986). Although most of the studies of ion balance in mangrove propagules and seedlings have focused on the more numerous viviparous and cryptoviviparous species, seedlings of the non-viviparous *Acanthus* also show somewhat lower sodium and chloride concentrations than adult plants. Again, salt appears to be partially excluded from the developing propagule, in this case a seed (Joshi *et al.* 1972).

Fecundity and parental investment

Avicennia in Costa Rica has been estimated to produce more than 2 000 000 propagules ha^{-1} year^{-1} (Jimenez 1990, quoted in Smith 1992). Translating this into a proportion of total net primary productivity is not easy. Collection of the total litter fall from mangrove trees and its separation into reproductive and other components indicates that the relative effort put into reproductive struc-

Fig. 1.14 Types of propagule and their properties in different mangrove species. Thin lines: non-viviparous, medium lines: cryptovivipary, thick lines: vivipary. (From data in Tomlinson, P.B. (1986). *The botany of mangroves*. Cambridge University Press.)

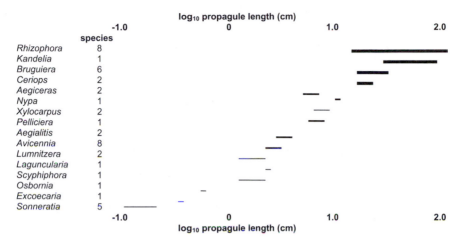

tures varies widely, according to species and location, but in *Avicennia*, *Rhizophora* and *Ceriops* typically ranges between 10 and 40 per cent (data in Bunt 1995). Vivipary and cryptovivipary therefore represent a considerable parental investment.

The parent tree may be able to modulate its investment. In *Avicennia marina*, only about 21 per cent of flower buds survive to form immature fruit: subsequent mortality means that only 2.9 per cent actually become viable propagules (Fig. 1.15). Similarly low survival rates have been estimated in *Rhizophora mangle* (07.2 per cent) and *R. apiculata* (13 per cent). Some of the decline in numbers in *Avicennia* can be ascribed to insect attack, but around 75 per cent of the mortality appears to be due to maternal regulation (Clarke 1992; Hutchings and Saenger 1987).

Dispersal

The long, pointed appearance of *Rhizophora* propagules hanging on the parent tree has led to the belief that they plummet like darts into the mud below and so immediately establish themselves. This may sometimes happen, but is probably rare. Propagules are often curved, and in any case are not well designed for flight, even vertically downwards. In addition, they may bounce off foliage or aerial roots, or land in water, or on mud that is too hard to penetrate. Besides, this mode of distribution would also result in virtually all seedlings attempting to establish themselves in the shadow of their parent. This is not the most favourable location: some degree of dispersal away from the parent is desirable.

The reality is more complex. *Rhizophora* propagules generally float for some time before rooting themselves. Initially, floating is horizontal. Over a period of

Fig. 1.15 Survival of buds, flowers and fruits of *Avicennia marina* before dispersal (N = 17 917). Error bars show standard deviations. (Reproduced with permission from Clarke, P.J. (1992). Predispersal mortality and fecundity in the grey mangrove (*Avicennia marina*) in south-eastern Australia. *Australian Journal of Ecology*, 17, 1618. Blackwell Science Pty. Ltd.)

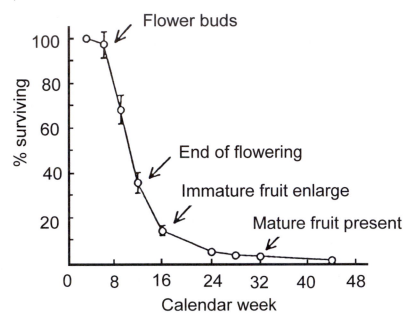

a month or so they shift to a vertical position. This makes it more likely that the tip will drag in the mud surface and result in the propagule stranding when the tide recedes. Roots first appear after 10 days or so, and many of the propagules lose buoyancy and sink. Presumably before this has happened the propagule is not ready to establish itself as a seedling. By 40 days, virtually all propagules show root growth (Banus and Kolehmainen 1975). Most will strand in a horizontal position, and erect themselves after rooting in the mud. Figure 1.16 shows the process of erection in *Avicennia*.

Propagules which do not successfully root after 30 days or so may regain buoyancy and float off again in a horizontal position. They may remain viable for a year or more (Rabinowitz 1978*b*). Occasionally, propagules are still viable after being transported tens of kilometres inland by hurricanes.

Other mangrove species show broadly similar patterns with a period of flotation, followed by sinking, rooting, stranding and establishment as seedlings. An exception appears to be *Avicennia germinans* of Central America, which does not sink and requires to be stranded before establishing itself. This is an interesting contrast with *A. marina*, the widespread Indo-Pacific species, which sinks on shedding its pericarp (Rabinowitz 1978*b*; Steinke 1975). Details of the propagules of a number of species are given in Table 1.6.

Fig. 1.16 Rooting and erection of a seedling of *Avicennia*. The initial root growth raises the body of the seedling, minimising contact with the anoxic mud. (Reproduced with permission from Chapman, V.J. (1976). *Mangrove vegetation*. Vaduz, J.Cramer.)

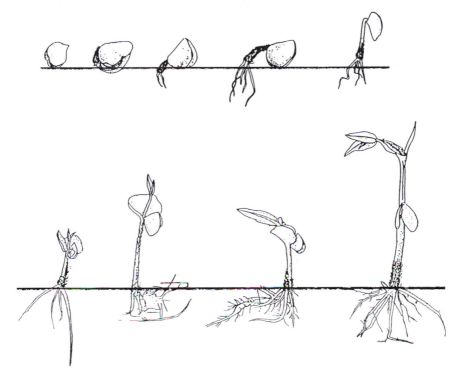

Table 1.6 Characteristics of propagules of a number of Central American mangrove species, data from Rabinowitz (1978*a*)

Species	Weight (g)	Length (cm)	Floating (days)	Obligate dispersal (days)	Longevity (days)
Laguncularia racemosa	0.4	2.1	23F, 31S	8F	35S
Avicennia spp.	1.1	1.83	always	ca 14 F/S	110S
Rhizophora mangle	14.0	22.1			
R. harrisonii	32.3	36.3	20–100F/S	ca 40 F/S	>365
Pelliciera rhizophora	86.2	7.8	1F, 6S	9F, 30S	70S

F, S = fresh, salt water conditions. *Avicennia* includes *A. germinans* and *A. bicolor*, virtually indistinguishable as propagules; weight and length of *Laguncularia racemosa* represent the means of Atlantic and Pacific populations; dispersal times of *Avicennia*, *Rhizophora* and *Pelliciera* are times at which 50% had produced roots.

The timing of these events is affected by circumstances. In sunny conditions, virtually all floating *Rhizophora* propagules pivot to the vertical by 30 days and root within a further 10 days or so; about half of shaded propagules are still floating horizontally after several months. This behaviour will facilitate settling in forest clearings rather than directly under adult trees (Banus and Kolehmainen 1975).

Salinity also affects dispersal. In several species, floating is prolonged in salt water compared with fresh, while *Avicennia* propagules shed their pericarps and sink more rapidly in brackish water (50 per cent sea water) than in either salt or fresh water (Rabinowitz 1978*b*; Steinke 1975). This presumably makes it more likely that settlement will be in a favourable environment.

Immediately after establishment, mangrove seedlings are tolerant of high salinity levels. In *Avicennia*, growth and development are maximal in 50 per cent sea water. Once the seedlings are mature and their reserves are depleted, the optimal salinity is 10 to 25 per cent sea water (Ball 1988*b*).

Why vivipary?

Vivipary (including cryptovivipary) is a rare phenomenon among the higher plants: mangroves apart, it occurs largely among plants of tropical shallow marine habitats, such as seagrasses (Elmqvist and Cox 1996). Because many viviparous mangrove species belong to a number of different families, whose other members are not viviparous, vivipary must have arisen through convergent evolution, on a number of separate occasions. This suggests some particular advantage of being viviparous in a tropical tidal, or shallow marine, environment.

It is often implied that the large propagules of many mangrove species enable widespread dispersal of offspring to colonise new areas. Certainly propagules that can float, and remain viable, for periods of weeks or months could in principle travel considerable distances. But do they?

The evidence is sparse. Approximately 78 per cent of *Avicennia marina* propagules (which sink after only a few days) were stranded within 2 km of their starting point, most of these within 500 m. With *Ceriops*, at least 75 per cent remained within 1 m of the parent tree, and 91 per cent within 3 m. In contrast, only 2 per cent of *Kandelia* propagules remained under the parent tree, 12 per cent within 50 m, and the remaining 88 per cent were not recovered and presumed to have dispersed greater distances (Chapman 1976; Clarke 1993; McGuinness 1997). Dispersal ranges probably vary greatly, depending on whether the propagules sink after a few days (*Avicennia marina*) or not (*A. germinans*), or even remain afloat and viable after months in sea water (*Rhizophora*) (Table 1.6).

Overall, however, there is little evidence that mangrove propagules regularly disperse over large distances in significant numbers. Nor, in fact, is there necessarily any advantage in scattering offspring thinly across large areas of sea. The chances of a propagule surviving the odyssey, landing in a suitable habitat, and successfully establishing itself there, must be remote compared to its prospects in the habitat whence it came. The chances are even more remote of two such pioneers settling close enough to have a chance of mutual pollination. The relationship of dispersal to the geographical distribution of mangroves is discussed further in Chapter 7.

It seems more probable that vivipary (and the production of large propagules in general) is an adaptation to the local habitat rather than a stake in the lottery of dispersal. On leaving the parental tree, a mangrove propagule must be physically able to withstand a certain amount of trundling around in the tide and currents: size may be an advantage here. A large propagule which sinks after a few days is likely to have travelled only a short distance, which probably means that it is still within the area in which its parent flourished, which must therefore be reasonably favourable for its own survival, while the accelerated settling in sunlight helps to avoid establishment in too shady a position. Finally, the accumulated stores of starch within the hypocotyl mean that rooting, erection of the propagule into a vertical position and respiration and growth can take place initially without being limited by the rate of photosynthesis. Seedlings of species with large propagules survive better than those with small propagules (Fig. 1.17). It would be interesting to know whether differences in propagule mass and post-establishment survival within a species were similarly related (Rabinowitz 1978*c*).

Not all mangrove species show vivipary, and not all have large propagules. Presumably different strategies have been adopted. Perhaps a species with large

Fig. 1.17 Relationship between propagule weight and seedling survival (half-life) for four mangrove genera. Diamond: *Laguncularia*, circle: *Avicennia*, open square: *Rhizophora*, filled square: *Pelliciera*. (From data in Rabinowitz 1978*b*).

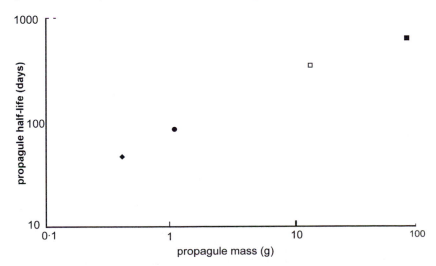

numbers of small, short-lived propagules exploits gaps in the forest that already exist, while the objective of the long-lived *Rhizophora* propagule is to be around when a new gap appears: dispersal in time, rather than space. Both strategies— and others intermediate on the spectrum between them—are therefore adaptations to an environment that is patchy in time as well as in space.

Why are mangroves tropical?

As mentioned on p. 2, mangroves are almost exclusively tropical. Geographical ranges are limited by either barriers to dispersal, restricted availability of suitable habitats, or by physiological constraints. Given the pantropical distribution of mangroves, and the numerous muddy estuaries in temperate regions, physiological constraints seem to be the most likely explanation for latitudinal limits on mangrove distribution.

The key mangrove adaptations are the ability to survive in waterlogged and anoxic soil, and the ability to tolerate salt or brackish water. Many, but not all, show vivipary and relatively large propagules. Why should this suite of characteristics be limited by latitude? Plants that tolerate salt, and plants that survive in waterlogged soils, are not unusual in temperate latitudes: in fact salt marsh vegetation is as typical of temperate coastal environments as mangroves are of tropical ones. There are temperate trees that cope with waterlogging, or with high salt concentrations. It is only the combination of being a tree, tolerating salt, and coping with waterlogging that is restricted to the tropics.

There are metabolic costs in tolerating salt, and in coping with waterlogging. Are there additional costs in growing and operating as a tree which mean that the mangrove way of life is not feasible where low temperature and short day length reduce photosynthesis for much of the year?

2 The mangrove ecosystem: can we see the wood for the trees?

The mangrove ecosystem can be considered at different spatial scales. At the level of the individual tree, there are complex adaptive structural and physiological responses to the immediate environment, particularly to physical factors such as salinity. These are discussed in Chapter 1. At a larger scale, local variations in these physical factors help to determine the overall structure of the mangrove forest. Such local variations, and the structure of the forest, are profoundly affected by the environmental setting in which it grows. Finally, the species present are a subset of those available in the geographical region, so over-riding biogeographical factors must be taken into account. These are considered in Chapter 7.

Any attempt at categorising is to some extent arbitrary and subjective. In the case of mangroves, classification is at best an attempt to impose order on 'a chaotic mess of special cases' (Thom 1984), and some oversimplification is inevitable. There have been many proposals for classifying the manifold forms of mangal, some of them applying quite well in certain parts of the world, but less well to geographically different situations (Ewel *et al.* 1998; Woodroffe 1992). This chapter discusses the general forms of mangrove forest, and the factors that shape them. The principal local factors are shore morphology, the influence of tides and river flow, and the variations in salinity and sedimentation that derive from these.

Form of the forest

Riverine mangroves

Many of the largest mangrove areas, particularly in Asia, are in the deltas of major rivers. Examples include the Indus delta of Pakistan (discussed further in Chapter 8), the Sundarbans where the Ganges and Brahmaputra rivers flow into the Bay of Bengal, the Merbok and Matang deltas of peninsular Malaysia, the Fly River of Papua New Guinea and, in South America, the mangroves of the Amazon delta. Characteristically, they have a low tidal range and strong freshwater flow carrying a considerable load of sediment, much of which is deposited in the mangroves. Before damming reduced the flow, the Indus, for instance, annually delivered more than 100 000 000 000 cubic metres of water and 200 000 000 tonnes of sediment to its delta (Chapter 8). Much of this was retained in the delta, rather than being flushed out to sea.

A typical large river delta comprises a number of distributaries splitting the land into many finger-like projections. Typically, the major channels shift to and fro as a result of silting, wave action and other factors, so that the currently active delta is often flanked by abandoned channels which may be penetrated by more saline water. Mangroves fringe these channels as well as the main distributaries and may indeed spread inland further up an abandoned channel than against the direction of flow of an active one. River-dominated settings therefore offer a range of different topographical and physical conditions in which mangroves may grow, and gradients in these which will affect species distributions as well as growth and productivity. It is also likely that the river flow will facilitate the outwelling to the sea of material generated within the mangal, and that a riverine mangrove will have an influence on other nearby habitats (Chapter 6).

Tide-dominated mangroves

Tide-dominated estuaries are characterised by a high tidal range over a shallow intertidal zone which can be colonised by mangroves. The tidal water is often full-strength sea water. Wave action is diffused by passage over the shelving intertidal zone. Typically, the river estuary is funnel-shaped, often with long-itudinally orientated intertidal shoals or mudbanks. Mangroves often appear as fringing forests along the edges of the estuary, although sometimes they may fringe open coast. Tides, like river water, can carry a considerable amount of sediment, which may be trapped within the mangal and contribute to the accumulation of sediment. Mangal sediment and other material can also be removed by tidal flows, so the flux of materials is likely to be bidirectional.

The seaward edges of tide-dominated mangals are often exposed also to con-siderable wave action. With relatively high wave activity and low river discharge the result is often a steeper shore. Wave action may rework sediment into barrier islands or sand spits enclosing broad lagoons. These are commoner in New World than in Indo-Pacific mangrove areas. Wave-dominated estuaries tend to be unstable, constantly changing in topography.

High river discharge and high wave action can combine to form a complex mixture of sand ridges, coastal lagoons, river mouths and abandoned distribu-taries providing a range of habitats for mangrove settlement.

Basin mangroves

On the landward side of fringing mangroves in estuaries are often large areas of mangrove, often termed basin or interior mangroves. These are sheltered from wave action, and often inundated only infrequently by the tides. Salinity varies greatly, depending on the circumstances. In areas of high rainfall, or of sub-stantial groundwater flow, salinity may be quite low. On the other hand, evaporation, and removal of water by the mangrove trees, may combine to raise the salinity, and in some areas the soil may be distinctly hypersaline, possibly as much as twice the salinity of sea water. Because currents, both riverine and tidal, are low and there is little turbulence, a basin forest is likely to represent a sink for nutrients and sediment, rather than a source for export (Chapter 6).

Riverine, tide-dominated and basin mangroves represent extremes of a continuum, rather than separate and distinct categories, and many mangals show elements of all three, merging into each other.

Special settings

In addition to the three major functional categories described above, there are some specific settings in which mangroves are found, which do not straightforwardly conform to this classification.

- **Drowned river valleys**: rising sea level has, particularly in parts of Australia, produced sheltered features known as rias, which are suitable for mangrove growth. A dramatic expansion in mangroves in the Holocene (8000–6000 years ago) followed a sea level rise (Chapter 8).

- **Overwash mangroves**: in the Caribbean, *Rhizophora* is commonly found in the form of small islets, completely overwashed by the tides (so usually with little leaf litter), and often underlain by an accumulation of peaty soil.

- **Carbonate settings**: on certain low-energy coasts in the tropics there are mangrove environments dominated, not by tidal or river flow, but instead by the accumulation of carbonate. Wave action may be dampened by a coral reef or sand barrier, behind which lime muds and peat (generated by the mangroves themselves) build up. In some cases the drowning of an old coral reef complex allows the growth of an irregular patchwork of mangroves in depressions in the underlying substrate. An example of this is seen in the mangroves of Ras Mohammed, in the Red Sea. These are at the northern limit of the geographical range of mangroves, and are also in the unusual position for mangroves of growing at the edge of the Sinai desert, a virtually unique combination of environmental circumstances.

- **Scrub mangroves**: these are usually found in extreme environments, where nutrients are limiting or little fresh water is available. Trees are stunted, and frequently no more than 1 m in height. Examples occur in Florida and in the Indus delta of Pakistan.

- **Hammock mangroves**: these are found particularly in the Florida Everglades. Relative isolation from rivers or the sea leads to a domed accumulation of organic peat over a depression.

- **Inland mangroves**: in some small islands of the Pacific and West Indies small pockets of mangroves are entirely cut off from the sea. These are usually in sink-holes or other depressions which may be either rocky or contain an accumulation of muddy sediments. In a few cases, local sea level changes have resulted in mangroves surviving in extraordinary situations. On Christmas Island, in the eastern Indian Ocean, a flourishing stand of *Bruguiera* has been found more than 20 m above sea level, where it may have survived in splendid isolation for 120 000 years (Woodroffe 1988).

Mangrove species zonation

Environmental settings are defined in terms of relatively large-scale land form and physical processes, while the functional classification approach shifts the emphasis to smaller spatial scale and to functional aspects of mangrove vegetation. Within these contexts, the different species of mangrove are not randomly scattered, but often occur in discrete and more or less monospecific zones.

Mangrove species zonation can be considered at different scales. On a tide-dominated shore, a clear vertical sequence of species often appears. An example is shown in Fig. 2.1. The reality is often not as simple as this stylised diagram would suggest: for example *Avicennia* often has a bimodal distribution, being abundant near the seaward margin and also some way upshore. A vertical sequence of species may repeat itself on the banks of tidal creeks and rivers, resulting in a complex two-dimensional pattern like that shown in Fig. 2.2, typical of mangrove areas of western Malaysia.

At a slightly larger scale, regular sequences of species may also occur with increasing distance up a river, interacting with vertical zones related to tide level. Even larger regional trends may emerge, such as the gradation of species across the vast delta of the Ganges and Brahmaputra, discussed later in this chapter (p. 42). Finally, of course, there are major geographical trends in the distribution of species, discussed in Chapter 7.

Fig. 2.1 Profile of a shore in north-eastern Australia (an area of high rainfall) to show mangrove zonation. HWS indicates the level of high water, spring tides. (Reproduced with permission of Dr N.C. Duke and the publishers from Tomlinson, P.B. (1986). *The botany of mangroves*. Cambridge University Press.)

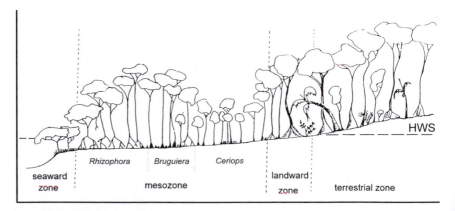

Fig. 2.2 Schematic plan of the complex pattern of species zonation typical of an estuarine mangrove area of western Peninsular Malaysia. (Reproduced with permission from Smith, T.J. (1992). Forest structure. In *Tropical mangrove ecosystems*. Coastal and Estuarine Studies no. 41, (ed. Robertson, A.I. and Alongi, D.M.), pp. 101–136, copyright by the American Geophysical Union. Adapted from Watson (1928). Mangrove forests of the Malay Peninsula. *Malayan Forest Records*, **6**, 1–275.)

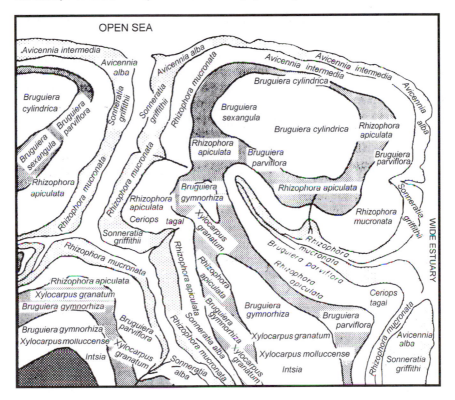

Not all mangrove areas show clear zonation patterns. Mangroves in Tanzania may be either zoned or unzoned, while those in neighbouring Mozambique are apparently unzoned. In the dwarf mangrove forests of south-eastern Florida the species are randomly intermingled, while in Australia classical zonation patterns appear to be the exception rather than the rule (Smith 1992; Snedaker 1982). The situation is therefore complex and whatever the underlying causes of zonation, they do not appear to apply universally.

Why do mangroves show zonation?

Before considering the possible causes of zonation patterns, it is worth remembering that a 'snapshot' view of adult species distribution may be misleading. Whatever the processes that promote zonation, they may operate differently on propagules, seedlings, saplings and adult trees. The distribution of adult trees is the culmination of all of the relevant processes that have affected all preceding life cycle stages. Additionally, measurement at a specific time, or over a short time period, of physical variables thought to be conducive to zonation may not give an adequate picture of the conditions that the adult trees have been subjected to throughout their lifetime. With these provisos, what can be inferred about the causes of mangrove species zonation?

Four main approaches to the causes of species zonation have emerged based on concepts from population dynamics, ecophysiology and geomorphology (Snedaker 1982).

The first suggestion is that zonation results from the physical sorting of floating propagules by water movement, the position of an adult tree within a forest being determined by the point at which a propagule was stranded. Secondly, selection might take place after settlement, for example by different species thriving at different positions along some physical gradient. This suggestion corresponds to the classical paradigm of shore zonation, developed largely on temperate rocky shores. Zonation might be a consequence of gradients created by major geomorphological changes. Finally, zonation might be the product of ecological interactions between species in the community, the sequence of species along a transect in space corresponding to a natural succession of species with time.

Propagule sorting

An ingenious explanation for mangrove zonation was put forward by Rabinowitz (1978a), stemming from her observation in Panama that the shore level at which a species occurred correlated with propagule size: the high shore species were those with smaller propagules. She suggested that the distribution of adult trees was based on tidal sorting of propagules. Larger propagules, such as those of *Rhizophora*, were more likely to be stranded and establish themselves at lower and more frequently flooded levels. In contrast, smaller propagules, for example those of *Avicennia* or *Aegiceras*, tended to be stranded on the upper shore where they had the necessary time to establish themselves between tidal inundations (Table 2.1; details of propagule size are also given in Fig. 1.14).

However, species found high on the shore are not always those with relatively small propagules. *Avicennia* and *Aegiceras*, for instance, are often abundant at low

Table 2.1 Shore level and mean propagule weight of species of mangrove in Panama, data from Rabinowitz (1978a)

Shore level	Species*	Mean propagule weight (g)
Lower	*Rhizophora harrisonii*	32.3 ± 2.6
	R. mangle	14.0 ± 1.1
	Pelliciera rhizophorae	86.2 ± 6.3
Higher	*Laguncularia racemosa*	0.41 ± 0.02
	Avicennia spp.	1.10 ± 0.11

Rhizophora harrisonii is found on the Pacific coast of Panama and *R. mangle* in the Atlantic. Propagules of the two *Avicennia* species (*A. germinans* (both coasts) and *A. bicolor* (Pacific)) were indistinguishable.

shore levels, as well as on the upper shore. Besides, irrespective of adult zonation, propagules of all species occur over virtually the entire tidal range. Differential dispersal of propagules is therefore unlikely wholly to explain species zonation, although it may play some part. Postsettlement differences in survival and growth are likely to be more important.

Physical gradients

The simplest explanation for zonation is that it represents a 'steady state' response to a gradient in some critical physical variable, and in the case of upshore/downshore gradients that this gradient is maintained by the tidal inundation regime. In many parts of the world, species distributions on a shore correlate well with tidal inundation regime. An example is the scheme developed by Watson (1928), based on Port Klang, on a mangrove river estuary in western Malaysia, but which has been more widely applied (Table 2.2).

Varying frequency and duration of submergence by the tides creates gradients in several physical variables, including salinity, to which mangroves might respond. Salinity would be an obvious candidate for a gradient determining species distribution because of its known effects on individual trees (Chapter 1). Each species, according to this view, would have an optimal salinity range and would be found at the level on the shore where this salinity prevailed. The relationship between shore level and salinity, of course, is not straightforward. In areas with considerable freshwater seepage or runoff, salinity will tend to increase in a seaward direction, while in an arid area with little or no freshwater from the land, and high evaporation rates, the reverse will be the case, and the high shore levels may be significantly hypersaline.

Table 2.2 Watson's tidal inundation classes, based on Port Klang, western Peninsular Malaysia, from Watson (1928)

Inundation class	Flooded by	Height above datum (feet)	Times flood per month	Species
1	all high tides	0–8	56–62	Rhizophora mucronata
2	medium high tides	8–11	45–59	R. mucronata Avicennia spp. Sonneratia
3	normal high tides	11–13	20–45	R. mucronata Ceriops candoleana R. apiculata Bruguiera parviflora
4	spring high tides	13–15	2–20	B. gymnorrhiza B. parviflora Lumnitzera littorea Acrostichum sp.
5	abnormal (equinoctial) tides	15	2	B. gymnorrhiza R. apiculata Acrostichum sp.

There are often (but not always) reasonable correlations between soil salinity, measured in the field, and mangrove species distributions. Some species, such as *Rhizophora mucronata*, are generally found only in relatively low salinities, below that of normal sea water. Others (for example *Avicennia marina*) can grow where the soil salinity is greater than 65 ppt (parts per thousand), compared with normal sea water whose salt concentration is typically around 35 ppt (Hutchings and Saenger 1987; Smith 1992).

Can we therefore conclude that salinity gradients are the principal determinant of species zonation? The hypothesis can be tested experimentally. If zonation results from physiological adaptation to different levels of salinity, then propagules transplanted to a range of different shore levels should succeed best in the same zone, at the same salinity, as their parental species.

The outcome of the experiment is not as predicted. Propagules of four species of mangrove were planted into both low and high intertidal forests differing in inundation frequency and salinity. Survival and growth were then monitored. All four species had better survival rates in the high forest, irrespective of where they had come from, and three out of the four species showed higher growth rates here. (The growth rate of the fourth species did not vary between sites). Only one of the species performed better at the level at which it was naturally most abundant; even so, it was out-performed here by the other three species (Smith 1987*a*). The results are summarised in Table 2.3. This does not indicate that salinity is irrelevant, but that it is not the key factor in mangrove species zonation. Other similar experiments lead to the same conclusion.

Laboratory experiments suggest that most mangrove species grow best at relatively low salinity. Rather than having different salinity optima, they differ in the range of salinity that they can tolerate. This is shown schematically in Fig. 2.3. The figure also indicates a further important feature of mangrove responses to salinity: the more euryhaline species (those that tolerate a wider salinity range) have in general lower growth rates *even at their optimum salinity*, than the stenohaline species which are restricted to a narrow range of salinity. The price of wide tolerance is slow growth. The result is zonation along a salinity gradient, because a slower-growing euryhaline species is outcompeted even where it would do best, and grows in less favourable salinities where more

Table 2.3 Growth and survival of propagules of four mangrove species at different shore levels, after Smith (1987*a*)

Species	Normal distribution	Survival best	Growth best
Avicennia marina	L	H	H
Bruguiera gymnorrhiza	L	L/H	H
Ceriops australis	H	H	H
Rhizophora stylosa	L	H	H

L = low tidal forest (high inundation frequency, low salinity), H = high tidal forest (low inundation frequency, high salinity).

Fig. 2.3 Ecological and physiological responses to a salinity gradient. The upper figure shows the distribution of species A, B and C along a salinity gradient, with each species dominant over a different salinity range. The lower figure shows the physiological responses of the same species to the salinity gradient. All three species have similar optima, but differ in their growth rate and range of tolerance. The result is that species A dominates at salinities around the optimum by out-competing its rivals. Species B and C are dominant at salinities which are sub-optimal for their growth but which prevent or limit the growth of other species. (Reproduced with permission from Ball, M.C. (1988). Ecophysiology of mangroves. *Trees*, **2**, 129–42, copyright Springer-Verlag GmbH.)

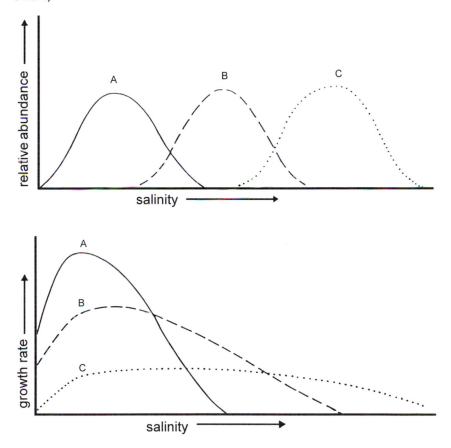

rapidly-growing species are at a relative disadvantage. Zonation results from an interaction between competition and salinity.

Although there are numerous accounts of zonation as a response to physical gradients, one question is rarely raised, probably because no simple answer is apparent. Gradients vary gradually and continuously: why, then, do species form distinct bands with often sharp boundaries? This may also be an effect of interspecific competition and mutual exclusion from the interface where that competition is strongest.

Salinity gradients were taken as an example. Other physical gradients could also contribute. Mangroves differ in their response to variation in other physical factors, such as soil anoxia (Chapter 1, p. 4), soil pH or sediment composition. Where these factors vary in line with salinity (as is often the case) a consistent pattern of physically-determined zonation is more likely to emerge. Where they vary independently of each other the outcome, in terms of species distribution patterns, may be more complex. This will be particularly true where there are interactions between physical variables. Mangroves appear to make trade-offs in their physiological responses to soil waterlogging and salinity (p. 18), and respond differently to salinity when soil textures differ (p. 14).

Geomorphological change

Shores consisting of soft sediment are dynamic. Erosion by current or wave action opposes accretion by sediment deposition: if these balance each other, a shoreline may be stable. If they are not in equilibrium, the shore either recedes or extends. Within a delta, distributary channels are apt to shift their course and sometimes become blocked and are replaced by new channels. All of these factors will affect physical gradients and any aspects of mangrove distribution that are affected by them.

In addition, the sea level may rise or fall. This may have either local or global causes. Locally, compaction and consequent subsidence of sediments has the effect of raising the level of the sea relative to that of the mangroves. At a larger scale, tectonic movements may raise or lower land levels. The inland mangroves of Christmas Island, now 20 m above current sea level, are an example (see above, p. 36). Another example is seen in the mangroves of the Sundarbans, at the north end of the Bay of Bengal. Tectonic action has raised the north-western part. Subsidence, due mainly to sediment compaction, has lowered it in the east. The resultant tilting, coupled with the massive input of fresh water in the delta of the Ganges and Brahmaputra to the east, has produced an overall salinity gradient, with the western delta much more saline. The result is an east–west gradient in the distribution of mangrove species, with less salt-tolerant species such as *Heritiera*, *Sonneratia* and *Nypa* being commoner in the east. This east–west gradient in species distribution interacts with the pattern of zonation parallel to the shoreline (Blasco *et al.* 1996).

Global changes in the amount of water in the oceans of the world also alter sea level locally. Such eustatic sea level changes affect the nature of mangrove zonation. The global impact of sea level change on mangrove communities is discussed further in Chapter 8.

If change is slow, mangrove communities will have time to adjust, and zonation and other distributions will reflect current conditions. If, however, the change takes place more rapidly, over a period of the same order as the lifespan of individual trees, an observed zonation pattern may have resulted from conditions that no longer obtain. Interpreting patterns in relation to current conditions may be misleading. Figure 2.4 gives an idea of the time scale of relevant geomorphological processes, and of the features of mangroves with which they might interact (Woodroffe 1992).

Plant succession

An alternative interpretation of the zonation pattern of mangrove forests is that the zones represent stages in a successional sequence, from pioneer species colonising the seaward fringe to mature climax forest behind. The sequence in space thus reflects an actual or potential sequence in time. Implicit in this model is the idea that each species so modifies the environment as to make it possible for a successor species to settle and, in due course, to displace its predecessor. In its extreme form, this theory envisaged mangroves actually creating land: pioneer species trapping mud, consolidating and extending the shoreline and virtually marching out to sea leaving behind them a trail of successor species culminating in a climax terrestrial forest. Specifically, *Rhizophora mangle* in Florida was credited with the creation of some 600 ha of land over a period of 30 to 40 years (Snedaker 1982). This successional progress was thought to be countered by disturbances, such as storms (see Chapter 8).

The idea that mangroves create land is now discredited (although it may be unfair to conclude that it is merely 'part of arm-chair musings of air-crammed minds' (Egler, quoted in Smith 1992). Mangroves undoubtedly trap sediment (see p. 47). In so doing, they may accelerate accretion, or retard erosion, of a shoreline. Many mangrove shorelines have mudflats on the seaward side of the

Fig. 2.4 Approximate time scale over which various physical and biological processes operate. (Reproduced with permission from Woodroffe, C. (1992). Mangrove sediments and geomorphology. In *Tropical mangrove ecosystems*. Coastal and Estuarine Studies no. 41, (ed. A.I. Robertson, and D.M.Alongi), pp. 7–41, copyright by American Geophysical Union.)

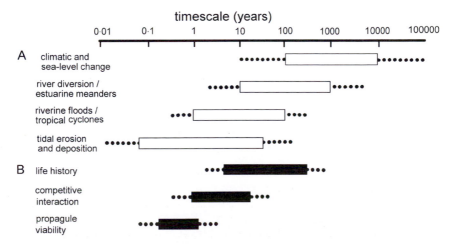

mangrove belt: here land has been created by sediment deposition without any part being played by mangroves. Conversely, it is not unusual to find mangrove trees literally falling out of an eroding shore. Accretion can be far faster than mangroves can exploit; and erosion faster than they can counter. It is much more likely to be the case that mangroves opportunistically follow a shoreline that is accreting for physical reasons, or retreat as a shore erodes than that they make the running.

Another problem with the idea of a seaward succession is that zonation patterns vary a great deal from place to place, and may show inconsistencies such as the bimodal distribution of *Avicennia* (see above and Chapter 4). This is hard to reconcile with the implication that zonation patterns simply recapitulate successional sequences.

This does not mean, of course, that mangrove forests do not show patterns of succession, only that succession is not the principal basis of zonation. Perhaps the most convincing account of succession comes from a remarkable long term study of the mature mangrove forest of Matang, western Malaysia. The Matang is a textbook example of a sustainably managed *Rhizophora apiculata* forest, and is discussed more fully in Chapter 8. Within the Matang area, plots were set aside by the Forestry Department in 1920 in order to determine growth and recruitment rates of trees under different thinning regimes; since 1950 every tree has been tagged and regularly measured, and natural changes in the plots monitored.

Since 1950, the pattern that has emerged is of occasional tree death leaving gaps in the canopy, and of these gaps being rapidly invaded by small plants such as the mangrove fern *Acrostichum*, the woody vine *Derris* and the shrub *Acanthus*. *Acrostichum* has a particular penchant for growing on the mounds thrown up by the mud lobster *Thalassina* (p. 105). These shrubby colonisers are followed by species of *Bruguiera*. *Bruguiera parviflora* has small and easily dispersed propagules which can form dense seedling patches under a *Rhizophora* canopy: further growth is suppressed unless a canopy gap appears. Subsequently *B. gymnorrhiza*, a slower-growing and longer-lived, shade-tolerant species, becomes more abundant. Overall, species richness has increased with time, typical of woodland successional sequences (Putz and Chan 1986).

There has also been a tendency for shade-tolerant species to increase with time. The first invaders of a patch are more likely to be those that do less well beneath the forest canopy and which need an open space to acquire enough light. Shade intolerance may also explain vertical zonation patterns: at the seaward fringe of a dense mangrove forest there is often a narrow zone of *Avicennia marina*. This species is less tolerant of shade than many of its competitors, and its presence at the edge of the forest may reflect this, rather than a preference for a low shore position (Smith 1987*b*).

The structure of a mangrove forest is therefore in part explainable in terms of 'patch dynamics': of gaps appearing by chance and being filled by a changing assemblage of species differing in composition (at least for a time) from the surrounding forest. Eventually something similar to the surrounding forest emerges. With a high incidence of gaps, a mangrove forest could be seen as a mosaic of patches of different successional age: if patches appear relatively rarely, the effect would be of transient aberrations in an otherwise homogeneous, or consistently zoned, environment.

The distribution of species that is actually found on a particular shore is therefore the outcome of an interplay between biotic and abiotic factors.

How different are mangroves from other forests?

Succession is part of the normal dynamics of many forest types: the chance appearance of gaps, rapidly colonised by opportunistic 'weeds' which are progressively ousted by slower-growing but more competitive species until a mature canopy forest reappears. Are mangrove habitats, then, just the same as other tropical forests (only wetter)?

Mangrove forests may in general be more subject to disturbance than other forests. As well as the normal disruptions that forests are prone to, mangrove forests may have to contend with typhoons, coastal erosion, wildly fluctuating river discharges and the tendency of delta channels to wander erratically to and fro. These fluctuations are superimposed on the more or less constant and predictable stresses of high salinity and soil anoxia. These, too, retard or reverse successional changes. Frequent change, coupled with environmental stresses, make it less likely that the processes of succession will culminate in a recognisable climax community (Lugo 1980).

Some of the differences and similarities between mangroves and their non-mangrove counterparts are shown in Table 2.4. The comparisons suggest that mangroves resemble (r-selected) pioneer species in their reproductive characteristics, but as adult trees they behave more as mature-phase competitive (K-selected) species. This observation, that mangroves contrive to have their cake and eat it (Tomlinson 1986) should prove a fruitful insight into the dynamics of mangrove forests.

Mangrove forest, compared with neighbouring tropical rain forest, is poor in species. This would be expected from an ecosystem in which the processes of succession—which include species accumulation—are inhibited. Even allowing for this, the number of plant species is low overall, and even lower locally, given the tendency for mangroves to grow in relatively homogeneous stands within a forest.

One reason may be that the species that occur in a particular mangrove forest must be a subset of the species of mangrove found in the region. The total number of species in a regional species pool, even in relatively species-rich regions such as South-east Asia, is not large. (By the standards of tropical rain forest, even the *global* pool of mangrove species is pitifully small.) If species are not present in the

Table 2.4 Comparison of the features of mangrove trees and community with pioneer and mature phase terrestrial forest communities, modified from Tomlinson (1986) and Smith (1992)

	Pioneer	Mature	Mangrove
Character			
Propagule size	small	large	variable (often large)
Propagule number	high	low	often high
Dispersal agent	often abiotic	usually biotic	abiotic (water)
Dispersibility	wide	limited	wide
Seed production	continuous	discontinuous	sometimes continuous
Seed dormancy and viability	long	short	variable
Seedlings	light-demanding not dependent on seed reserves	not light-demanding dependent on seed reserves	light-demanding? dependent on reserves
Reproductive age	early	late	early
Life span	short	long	long?
Leaf size	often large	medium/small	medium
Leaf palatability	high	low	low
Wood	soft, light	hard, heavy	hard, heavy
Crown shape	uniform	varied	uniform
Competitiveness	for light	for many resources	mainly for light
Pollinators	not specific	highly specific	not specific
Flowering period	prolonged or continuous	short	prolonged or continuous
Breeding mechanism	usually inbreeding	usually outbreeding	usually inbreeding
Community			
Species richness	low	high	low
Stratification	low	high	low/absent
Size distribution	even	uneven	even?
Large stems	absent	present	absent (except on old undisturbed stands)
Undergrowth	dense	sparse	usually absent
Climbers	few	many	few
Epiphytes	few	many	few

geographical area, they cannot be present in a particular mangrove forest. The biogeography of mangroves is discussed further in Chapter 7.

Mangroves differ from other forests in one other marked respect: they virtually lack any understorey vegetation (Fig. 1.4). There is often, in practice, a patchy understorey of mangrove seedlings, but only rarely a significant development of understorey vegetation of other species. Almost the only exception to this is the mangrove fern *Acrostichum*, but even this is not a true understorey plant, as it depends heavily on patches of direct sunlight. Possibly shrubs cannot grow because of the combination of low light levels under the forest canopy, high salinity and anoxic soil. Whatever the reason, the lack of understorey vegetation, and the relatively low tree species diversity, mean that mangrove forests are relatively simple in physical structure (Janzen 1985). Physical complexity also tends to reduce as salinity increases (Brown and Lugo 1982).

Mangrove forests therefore are not just tropical forests that happen to live in salt water and anaerobic soil. They have special features of their own: relatively few species and a simple physical structure, and marked zonation as the consequence either of a frustrated successional sequence, of the effects of geomorphological change, of physiological responses to clinal physical variables. It goes without saying that these are not mutually exclusive.

The mangrove environment: do mangroves make mud?

The distribution and growth patterns of mangroves can largely be interpreted as physiological responses to variations in their physical environment, modulated by competitive interactions between the trees themselves. As the physical environment affects the trees, so do they in turn alter their environment. A mangrove forest is not just a mudflat in which trees happen to be growing: the trees affect their substrate. Do they also create it?

The question of whether mangroves create land has already been briefly mentioned (p. 43). Although this appears not to be so, there is evidence that established mangrove forests do trap sediment particles, so may accelerate accretion. Conversely, their ability to trap sediment may retard erosion.

The dynamics of sediment in mangrove forests has been studied in only a few cases, and is far from being clearly understood. Tidal currents, wave action, river flow, salinity gradients and the topography of mangrove habitats all interact and affect sedimentation in various complex ways.

Many mangrove areas are estuarine, and most large rivers carry a considerable sediment load (see p. 34 above). As the river water passes through its mangrove estuary, the sediment it carries either settles or remains in suspension and is carried out to sea. A key variable in deciding the outcome is settling velocity, the rate at which particles will sink in still water. If the settling velocity is low in relation to current speed and turbulence, particles will remain suspended.

In fresh water, the settling velocity of fine suspended clay particles is low. In the Fly River estuary, a mangrove area in Papua New Guinea, for instance, settling velocity is typically between 10^2 and 10^1 cm s^1. However, when fresh water mixes with salt in the estuary, the suspended particles start to flocculate and form larger aggregates whose settling velocity may be an order of magnitude higher. Sometimes small planktonic animals become trapped by flocs and carried to the mud surface. The process of flocculation can start at salinities as low as 1 part per thousand, about 3 per cent of the concentration of normal sea water. The reasons for flocculation are complex, and relate to the effects of ions on the electrostatic charge on the particle surface, as well as to biological activity: the effect, however, is to promote sinking. The effect may be locally enhanced by salt exuded from mangrove pneumatophores (Wolanski 1995; Young and Harvey 1996).

Suspended particles are also carried in incoming tidal water, and tidal sediment can also be trapped in the mangrove forest. Here, too, the process is complicated and far from fully understood, and several factors are involved. Tidal currents move rapidly up tidal creeks, then spread out over the mangrove forest floor. The flow rate is much greater in the creek than in the forest: in one mangrove system in northern Australia, current velocity was measured in both creeks and forest. In the former it was typically greater than 1 m s^{-1}, while in the latter it seldom rose above 0.1 m s^{-1}. There are two main reasons for this. In the first place, the area of forest may be large in relation to the creek, hence the water moves more slowly when it spreads out. Secondly, and more significantly, the forest offers resistance to water flow. Tree trunks, roots and pneumatophores add to the friction already created by the mud surface.

Small-scale turbulence around roots keeps particles in suspension when the tide is advancing relatively fast. As the tide advances, particles are carried upshore. At slack water high tide, these particles can sink. The retreating tidal current is too low to resuspend and remove them. The result is net accumulation of sediment from the sea: some 80 per cent of suspended sediment brought in from coastal waters is trapped in mangroves in this way (Furukawa and Wolanski 1996; Furukawa *et al.* 1997).

Net sedimentation rates in mangrove forests vary: measurements of up to 8 mm year^{-1} have been made, although most estimates are lower than this (Woodroffe 1992). Sedimentation rate correlates with pneumatophore density in an *Avicennia* forest. It also correlates with the density of simulated pneumatophores, apple tree cuttings of appropriate size planted in bare mud at various densities (Young and Harvey 1996). This confirms that mangrove trees do increase the rate of sedimentation, and thus actively contribute to their environment.

As sediment builds up round *Avicennia* pneumatophores, gas exchange and root respiration will be progressively restricted. This tendency may eventually lead to the death of the tree and either soil erosion or, possibly, its replacement by a successor species. Alternatively, it may be that *Avicennia* can adjust pneumatophore growth to balance the enhanced sedimentation (Blasco *et al.* 1996; Young and Harvey 1996).

Mud is of course a crucial component of the mangrove ecosystem. The mud surface may be the site of significant photosynthesis by unicellular algae and bluegreen bacteria. Below the surface bacteria and fungi decompose the organic components of the mud. Many animals also burrow beneath the surface: these may or may not emerge on the surface at low tide. Mud-dwelling animals include a host of deposit or filter-feeders, detritivores, herbivores and predators. They are discussed in Chapter 4.

As well as contributing to the mud, mangrove trees provide a hard substrate on which other organisms can grow. Pneumatophores, aerial roots and even the lower branches and leaves are often festooned with algae, or covered with barnacles and oysters. These in turn are fed on by a host of predators.

Parts of the mangrove trees that lie beyond the reach of the tides represent an environment not greatly different from that of a terrestrial forest, occupied by an essentially terrestrial fauna of insects, mammals and birds. These invaders from the land are described in Chapter 3.

Finally, of course, mangroves are a major source of net primary productivity. Particularly in the form of leaf litter, but also reproductive products, twigs and eventually entire dead trees, this dominates energy flow pathways, directly or indirectly fuelling virtually the entire ecosystem. The contribution of mangrove trees to the structural complexity and productivity of their environment is obvious. They also affect the environment more subtly. The problems of maintaining aerobic roots in anoxic mud were discussed in Chapter 1. Sub-surface transfer of oxygen, by means of aerial roots and pneumatophores, is so effective that the mud in the vicinity of underground mangrove roots is less anoxic than that at some distance from the root. Mangroves roots oxygenate their environment (p.10).

Mangroves also have local effects on soil salinity. Figure 2.5 shows the salt concentration along a transect through a belt of mangrove trees and adjoining salt flat in northern Australia. Salinity in the soil under the mangroves is constant at around 50 ppt, but immediately outside the mangrove belt it rises sharply. It seems that the presence of mangrove trees prevents extreme hypersalinity. This could be because they remove salt from the soil and get rid of it by salt gland secretion (see p. 13) or by shedding senescent leaves containing accumulated salt. Alternatively, evaporation (and concentration of residual salt) from the mud surface may be less in the shade of mangrove trees than on open

Fig. 2.5 Salt concentration profile for a transect through a mangrove and adjacent mudflat. Salt concentrations are in g l^{-1}. (Reproduced with permission from Ridd, P.V. and Samm, R. (1996). Profiling groundwater salt concentrations in mangrove swamps and tropical salt flats. *Estuarine, Coastal and Shelf Science*, **43**, 627–35, copyright Academic Press.)

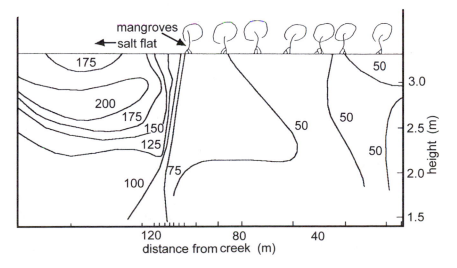

Fig. 2.6 Variation of soil ammonium concentrations with depth during a season of active growth (a: September) and of little growth (b: April) to demonstrate depletion of ammonium by mangroves during the period of active growth. (Reproduced with permission from Boto, K.G. and Wellington, J.T. (1984). Soil characteristics and nutrient status in a northern Australian mangrove. *Estuaries*, 7, 61–9, Estuarine Research Federation.)

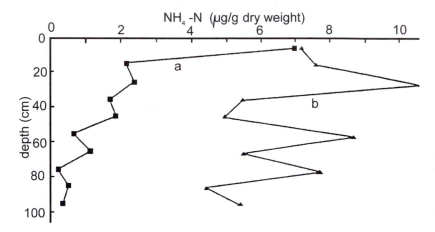

mud. A final possibility is that mangrove soil is permeated by crab burrows (Chapter 4), and water flowing through these may remove high salinity water from the area. (Ridd and Sam 1996).

At a smaller scale, water is continuously being taken up through the mangrove roots and, in species such as *Avicennia* and *Aegiceras*, much of the salt is excluded at the root surface. This leads to accumulation of salt just outside the root. This very local increase in salt concentration may reach such proportions as to limit the growth rate of mangrove seedlings (Passioura *et al.* 1992).

Given the considerable biomass of a typical mangrove forest, and the demand for nutrients such as nitrate and phosphate (p. 19), it is obvious that mangroves will have an effect on prevailing nutrient levels. Evidence that mangroves are nutrient-limited circumstantially confirms this. Direct evidence comes from comparison of soil ammonium levels in a mangrove forest at a time when the trees are actively growing (and requiring nutrients) with levels at a time of inactivity. As Fig. 2.6 shows, active growth greatly depletes soil ammonium levels.

Mangrove trees transform what might otherwise be relatively uniform mud into a structurally complex and highly productive environment, and affect levels within that environment of oxygen, salt, and inorganic nutrients. A mangrove habitat is a great deal more than mud which happens to have trees growing in it.

3   The mangrove community: terrestrial components

A mangrove community is more than just an assemblage of trees physiologically adapted to living in brackish water. Living in, on, or around the mangrove trees is a heterogeneous community of organisms, which depend on them for attachment, shelter or nutrients. The mangrove trees may suffer or benefit from their presence. They may be permanent residents, or may occupy the mangal temporarily, either seasonally or for part of their life cycles. They may occur exclusively in mangrove habitats or be found in any humid tropical environment. In addition, vast numbers of bacteria and fungi are involved in decomposition of detritus. Although little is known in detail about the species involved, their function within the ecosystem is crucial, and is discussed in Chapter 5.

Mangrove-associated animals or plants are of either terrestrial or marine origin. Firstly, we shall consider the terrestrial components of the mangrove community.

Mangrove-associated plants

Mangrove trees provide a firm substrate on which other plants can grow. Climbers include such plants as the vines *Caesalpina* and *Derris* (Leguminosae), lianes, the mangrove rattan palm (*Calamus erinaceus*) of Malaysia, the orchid *Vanda* and climbing ferns. These are rooted in the soil, usually on the landward fringe of the mangal, but use mangrove trees as support and may extend their shoots into the more seaward mangrove trees.

Some plants are truly epiphytic, and grow entirely on mangrove trees with no direct connection with the soil. These include lichens growing on the bark of trunks and branches, and vascular plants. Among the latter are several species of fern and orchid and, in the New World, bromeliads.

In general, the species of vascular plant associated with mangroves, whether as climbers or true epiphytes, are the same as those that occur in adjacent terrestrial communities. They are unable to tolerate high salt levels, and therefore do not penetrate deeply into the mangrove habitat. There are, however, some apparent exceptions. Some bromeliads, for instance, have succulent leaves and seem to accumulate salt within their tissues. This suggests that they may have evolved a degree of salt tolerance in parallel to the mangrove trees on which they grow (Gomez and Winkler 1991).

One group of plants which depend on mangroves for more than physical support are the parasitic mistletoes (Loranthaceae) of Australia and Papua New Guinea, and similar mistletoe species in other parts of the world. Mistletoes have virtually lost their root system and, instead, tap into the xylem of their host with a penetrating structure known as a haustorium. Because the salt concentration in mangrove sap is higher than that of terrestrial trees (Chapter 1), the tissues of the mistletoe must presumably be as salt tolerant as those of its host. This is probably relevant to the degree of host specificity shown by some mangrove mistletoes. *Amyema thalassium*, for instance, grows only on the mangroves *Avicennia*, *Excoecaria* and *Bruguiera*. Other species are more indiscriminate.

Another apparent specialisation is the close similarity between the leaves of the mistletoe and those of its host. Resemblance between parasite and host is common among parasitic plants. It may be due to mimicry, so that the plant is well concealed among its host's foliage. Many mangrove species are relatively unpalatable to herbivores, and mangrove mistletoes may, by their resemblance, deter potential herbivore attack without the cost of producing their own aversive substances. Alternatively, the factors that cause mangrove leaves to evolve their particular configuration may apply equally to the mistletoe: an example of convergent evolution, rather than mimicry. Mangrove species themselves, of different taxonomic origins, are markedly similar to each other in leaf structure, implying convergence. Apart from the fact that it uses its host's root system rather than its own, a mistletoe is just another mangrove species and it is not surprising that it resembles its colleagues in appearance.

A further peculiarity of mistletoes is their means of dispersal. In the case of the mangrove mistletoes, this is by the mistletoe bird (*Dicaeum hirundinacaeum*). These have a relatively short intestine, and the mistletoe berries appear to have laxative properties. The combined effect of these two features is that berries pass through the bird within 10 to 45 min, and travel only 3 to 5 km before being wiped off on branches, where they germinate (Barlow 1966; Hutchings and Saenger 1987).

A final group of epiphytes, more or less confined to mangroves, are the so-called ant-house plants. As their name suggests, these have a special relationship with ants. They are discussed later (p. 58).

Animals from the land

Apart from the special features discussed in Chapter 2, mangroves are not very different from other, non-mangrove forests in their vicinity. Indeed, mangroves often shade almost imperceptibly into adjacent habitats. It is not surprising, therefore, that a substantial proportion of the fauna of mangroves comprises species obviously derived from neighbouring terrestrial environments. Most of the major groups of terrestrial animals are significantly represented in mangroves.

Insects

As anyone who has worked in mangroves can testify, insects are generally present in abundance. Mangrove insects include herbivores, feeding on leaves, flowers, seeds or mangrove propagules; detritivores, eating dead wood or decaying leaves; more general foragers; and predators. Some insects play crucial roles as pollinators: these are discussed in Chapter 1. Finally, of course, insects in their turn represent a major food source for predators. Although mosquitoes and ants are often hard to ignore, other insects which do not make their presence felt to quite the same extent may be of even greater ecological significance. Surprisingly little is known of the significance of mangrove insects.

Insect herbivory

The most obvious signs of insect herbivory are on mangrove leaves, which often show erosion of the leaf margin, holes in the blade of the leaf or evidence of leaf-miners. Table 3.1 gives estimates of the percentage of leaf area eaten by insects in 14 species of mangrove surveyed at Missionary Bay, Queensland. Simple measurements of missing leaf area exaggerate the extent of the damage, since holes expand as leaves grow. Allowing for this, the proportion of leaf productivity claimed by insect herbivores in this case is probably, on average, only around 2.1 per cent (Robertson and Duke 1987). Other estimates, from Belize, China, Malaysia and other parts of the world support the general proposition that insect herbivory typically accounts for only a small proportion—generally under 5 per cent—of leaf production (Farnsworth and Ellison 1991; Lee 1991; personal observations).

Table 3.1 Estimated leaf damage by insects (percentage of leaves with damage and estimated leaf area removed) for 14 mangrove species at Missionary Bay, North Queensland, data from Robertson and Duke (1987)

Species	Damaged leaves (%)	Area loss (% ± s.e.)
Acrostichum speciosum	64	3.1 ± 0.3
Aegiceras corniculatum	99	16.0 ± 0.9
Avicennia marina	89	8.8 ± 0.6
Bruguiera gymnorrhiza	51	3.7 ± 0.4
Ceriops tagal	62	6.3 ± 0.4
Excoecaria agallocha	8	0.3 ± 0.7
Heritiera littoralis	100	35.0 ± 1.6
Lumnitzera littorea	73	4.3 ± 0.5
Rhizophora apiculata	63	5.8 ± 0.5
R. lamarcki	34	1.4 ± 0.2
R. stylosa	57	5.1 ± 0.4
Scyphiphora hydrophyllacea	12	0.7 ± 0.1
Xylocarpus australensis	18	3.0 ± 0.5
X. granatum	94	10.0 ± 0.7

Although insect attack on leaves may on average be low, the extent varies considerably, both within and between species. In Belize, for instance, leaf damage ranged from 4.3 to 25.3 per cent in *Rhizophora*, and 7.7 to 36.1 per cent in *Avicennia* (Farnsworth and Ellison 1991), while Table 3.1 shows that species vary between 0.3 per cent *(Excoecaria)* and 35 per cent damage *(Heritiera)*.

Why the variation? The answer lies in considering another question: what limits insect attack? Not all leaves are equally susceptible to attack. *Rhizophora* leaves are quite tough, even leathery, while those of *Aegiceras* and *Avicennia* are softer and more succulent; and younger leaves are probably physically easier to eat than older ones of the same species.

Leaves vary also in their chemical composition. Plants are not necessarily passive victims of herbivore attack. They produce a variety of substances that may be aversive to insect attack, actually toxic or in some other way reduce the value to the would-be attacker. Even salt secretion on mangrove leaves (Chapter 1) may deter some herbivorous insects such as coccids (Clay and Andersen 1996). More importantly, mangroves are often rich in soluble tannins, saponins or various secondary metabolites which inhibit herbivory (Chapter 4). *Rhizophora* leaves have higher tannin levels than *Avicennia*, and lower levels of insect attack. The species with the lowest level of insect damage is *Excoecaria agallocha*. When the leaves of this species are damaged, they produce a toxic latex which strongly irritates human skin and can be lethal to fish. It seems unlikely to be very beneficial to a grazing insect (Robertson 1991).

If plants evolve defences against insects, insects in turn can evolve tolerance of defensive chemicals. This is probably why the insect fauna differs between different mangrove species, a result of adaptation by different insect species to different suites of defensive chemicals (Farnsworth and Ellison 1991).

If the potential dangers of herbivory vary, so too do the benefits. Mangrove leaves vary in nutritional value. A simple index of this is the carbon: nitrogen (C: N) ratio. A leaf with a low C: N ratio is richer in nitrogen and of greater value to a herbivore. It is therefore predictable that insects would eat leaves of *Avicennia*, with a relatively low C: N ratio, in preference to those of *Rhizophora*. Insect herbivory is sometimes higher when leaf nitrogen levels have been raised by enrichment of the soil with bird droppings or artificially, although this has not always been found (Farnsworth and Ellison 1991; Feller 1995; Onuf *et al.* 1977).

The extent of herbivory therefore depends on a host of factors affecting the palatability and value of leaves, and these vary with the age of the leaf, probably seasonally and between species.

Whereas it is relatively easy to assess the extent of leaf damage, it is not always possible to identify the species responsible. Many species are involved: in the mangroves of Belize, more than 66 leaf-eating species have been identified, while the tally of herbivorous insects in the Andaman and Nicobar islands is nearly 200. The most important are probably lepidopteran caterpillars, although the depredations of chrysomelid and lampyrid beetles, and homopteran bugs, may also be significant (Farnsworth and Ellison 1991; Lee 1991; Veenakumari *et al.* 1997).

Occasionally, for unknown reasons, attack by herbivores may reach epidemic proportions. Even *Excoecaria*, with its toxic sap, is not exempt. In one outbreak, in Indonesia, every tree in an area of 5 to 10 km^2 was completely defoliated by caterpillars of the noctuid moth *Ophiusa* according to the local press 'as if by demons' (Whitten and Damanik 1986).

In the Mai Po marshes of Hong Kong severe defoliation of *Avicennia* by larvae of the pyralid moth *Nophopterix* is an annual event. Figure 3.1 shows the sequence of events. Over about 10 days, *Nophopterix* caterpillars increase from undetectable levels to a peak of more than one individual per leaf. Twenty days later, leaf damage has risen to nearly 80 per cent. It then declines due to the loss of old, damaged leaves and their replacement by new growth (Anderson and Lee 1995).

Such a dramatic event has a profound impact on the *Avicennia* population. Propagule production is very low, and young seedlings scarce. Very few trees reproduce successfully, except in years when defoliation is less severe than normal. The connection between the presence of caterpillars and failure to produce propagules was confirmed by preventing defoliation by treating the

Fig. 3.1 Densities of caterpillars of the pyralid moth *Nophopterix syntaractis* (open symbols) and leaf damage (filled symbols) throughout a defoliation event in mangroves of Hong Kong. (From Anderson, C. and Lee, S.Y. (1995). Defoliation of the mangrove *Avicennia marina* in Hong Kong: cause and consequences. *Biotropica*, **27**, 218–226, copyrighted 1995 by the Association for Tropical Biology, P.O. Box 1897, Lawrence, KS 66044–8897. Reprinted by permission.)

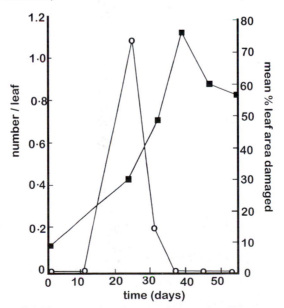

trees with insecticide. In this case flower and propagule production rose to more normal levels.

Defoliating caterpillars may therefore have a significant impact on productivity and reproduction of a mangrove forest. In this case, some 43 per cent of the original leaf biomass was removed, equivalent to 3.4 tonnes per hectare (t ha^{-1}). The consequent effect on reproductive rate affects the age structure of the population. Defoliation may, in the long term, threaten the dominance of *Avicennia* at Mai Po, since it will allow other species, whose seedlings thrive in unshaded conditions, to flourish. Finally, much of the leaf material removed is converted to caterpillar faeces and moultskins, and soluble nutrients leach out of this material more readily than from leaf litter. Nutrient cycling, or export, may therefore be facilitated by the caterpillars. Simple leaf eating may have profound effects on ecosystem structure and function (Anderson and Lee 1995).

Insects also attack developing flowers, fruit and propagules both before and after they leave the parent tree. Common culprits are borers such as the scolytid beetle *Coccotrypes*, species of which infest *Rhizophora* in the Caribbean and Australia, and moth larvae.

Losses occur throughout the development of buds, flowers, fruit and mature propagules (Fig. 1.15). In *Avicennia marina* studied in southern Australia, only a tiny minority of flower buds achieve their potential as propagules. Excluding insects increases fruit survival; nevertheless, most of this mortality is due to causes other than insect predation, such as regulation of propagule production by the parental tree (Clarke 1992).

Dispersed propagules frequently show signs of insect damage, which may have been initiated either before or after shedding of the propagule by the parent tree. In one survey, the proportion of damaged propagules ranged from 2.1 to 92.7 per cent, depending on species, location and year (Table 3.2). Recording

Table 3.2 Incidence of propagules damaged by insect attack (%) at various sites in tropical Australia (data from different sites separated by a comma, data from different years by a dash), from Robertson *et al.* (1990)

Species	% insect attack
Avicennia marina	59.1, 63.0, 64.8
Bruguiera exaristata	56.0
B. gymnorrhiza	11.7–44.4, 22.9
B. parviflora	33.6, 42.1, 48.2, 57.1
B. sexangula	43.6
Ceriops australis	6.2
C. tagal	7.2, 8.2
Heritiera littoralis	67.8–92.7
Rhizophora apiculata	3.1, 3.5, 6.3, 24.7
R. stylosa	2.1, 12.5, 33.5
Xylocarpus australoasicus	19.8, 68.4
X. granatum	59.1, 75.9

visible signs of damage probably underestimates actual losses: not all *Heritiera* seeds, for instance, showed visible signs of insect damage, but when examined more closely, only one out of 435 seeds actually contained an intact embryo. In other species, visible insect damage seemed to have no significant effect on propagule survival or seedling growth (Robertson *et al.* 1990).

Reproductive losses to insects must be seen in the context of massive losses from other causes. After all, if a developing embryo dies because of maternal controls, is eaten by a crab, uprooted by a monkey, or simply fails to find a suitable site in which to take root, the fact that it is also being bored by a beetle will not matter very much. Before assessing the overall importance of insect attack on mangrove reproductive capacity it will be necessary to know a great deal more about its incidence, and about other causes of mortality.

Termites

Termites are an important component of the fauna of tropical mangroves, but very little is known about them. For the most part, they burrow inside the trunks and branches of mangrove trees, particularly towards the landward fringe of the mangrove forest. An exception is a species common in Malaysia (possibly *Nasutitermes*), which constructs an external nest on a tree trunk several metres above the highest point reached by the tide, with narrow galleries snaking down the trunk to the aerial roots, and upwards to the canopy. Termites may be very important in destroying dead wood: a dead tree trunk, if poked, sometimes crumbles and collapses, revealing a mass of termites and little else left of the interior.

Ants

Ants are often abundant in mangrove trees, particularly in the canopy, as they are in other tropical forest vegetation. Not much is known about ants in mangrove forests, but their abundance suggests their ecological significance.

One conspicuous species is the weaver or tailor ant (*Oecophylla smaragdina*) found in mangroves of the Indo-Pacific. Captain Cook came across these in Australia, where they made a deep impression on him as 'a remarkable kind of ant . . . when the branches were disturbed they came out in great numbers, and punished the offender by a much sharper bite than ever we had felt from the same kind of animal before' (Quoted in Simberloff 1983). Weaver ants usually nest in *Bruguiera*, *Sonneratia* or *Ceriops* trees. The species gets its colloquial name from its nests, which are constructed by stitching folded leaves together using sticky secretions from a larva. *Oecophylla* feeds on the sugary secretions of coccid bugs which suck plant sap, supplemented by protein from preying on coccids and other insects. Other ant species, such as the leafcutter ant (*Atta*) of South America, feed directly on mangrove leaves.

Ants may therefore have both harmful and beneficial effects on their host trees. There is some evidence that the mangroves *Laguncularia* and *Conocarpus* of

Florida attract ants by means of extrafloral nectaries which secrete a sugary solution. Presumably the costs of the secretion are outweighed by the benefits of the protection conferred by the ants against herbivores. Captain Cook would confirm the deterrent effect of ants.

A more intimate dependency has been established between ants and certain species of plants epiphytic on mangroves, known as 'myrmecophytes', or ant plants. For instance, in Australian *Myrmecodia* plants, which may weigh several kilograms, have a bulbous stem honeycombed with tunnels occupied by the ant *Iridomyrmex* (and, in addition, a butterfly larva). Ants living in such 'ant-house' plants clearly gain protection: is there any advantage to the plant? Another myrmecophyte species, *Hydnophytum formicarium*, has specialised absorptive chambers. Ants deposit their debris here, and it has been demonstrated experimentally that when the colony is fed radioactively labelled *Drosophila* larvae radioactive compounds are absorbed into the plant. The relationship is therefore mutual: ants obtain shelter, and the plants a supply of scarce nutrients, particularly nitrogen. Saprophytic fungi growing in the ant galleries probably play a role in releasing soluble nutrients from the ant debris. To make the situation even more complicated, the ants also tend larvae of the butterfly *Hypochrysops* which feed on the tubers and leaves of the ant plant. An epiphytic plant therefore grows on a mangrove tree, accommodates ants, which tend butterfly larvae and supply nutrients to their host, aided by fungi: two plants, one or more fungi, and two animal species interacting (Huxley 1978; Janzen 1974).

Ants are essentially terrestrial animals. It is not surprising, therefore, that the ants found in mangroves are largely arboreal, and construct their nests well above the highest point reached by the tide. Most species are shared with neighbouring terrestrial forest. Some, such as the leafcutter *Atta*, may nest outside the intertidal zone and forage in the mangal only at low tide. There are, however, a few species of ant restricted to mangroves. In Australia, *Polyrachis sokolova* is found only on the lower shore, in the *Ceriops* and *Rhizophora* zones. Uniquely, it nests actually in the mangrove mud. Because of the position of the nests, they are inundated in up to 61 per cent of high tides, for periods of up to 3½ hours. Individual *P. sokolova* forage at low tide, returning to the nest before the entrances are covered by the tide. Examination of food remains in the nest shows a broad range of items: small decapod and amphipod crustacea, lepidopteran larvae and other insects. They have also been observed feeding on bird droppings.

Are these ants adapted to a truly aquatic life? They are able to run or swim on the water surface in order to reach the nest entrance before the tide covers it, but this is not truly aquatic behaviour. The hairy surface of the cuticle can hold an air film during submersion, but ants do not live long under water: at 33°C only 50 per cent survive 3½ hours immersion. The structure of the nests has been studied by making casts with polyurethane foam. They may be up to 45 cm deep in the mud, and have two entrance holes. When the advancing tide reaches the nest, loose soil collapses into the entrances, blocking them and preventing water from penetrating the passages beneath. In this way ants—provided they get back to the nest in time—survive high tide in air trapped in the nest galleries (Clay and Andersen 1996; Nielsen 1997).

Mosquitoes and other biting insects

The record for the number of mosquitoes attracted is probably held by William Macnae; over 80 settling on his bared arm within 2 minutes (Macnae 1968). Many workers in mangroves probably feel they have come close to this. Mosquitoes, and other biting insects, such as the 'sandflies', or biting midges (Ceratopogonidae), attract attention because of their nuisance value, and because in many cases they are vectors of diseases such as malaria and yellow fever.

To breed, mosquitoes require shallow and fairly stagnant pools of water. These occur in abundance in mangroves. Rot holes in trees, tidal pools, and the water retained in crab burrows all provide suitable environments for egg laying and larval development. Different species of mosquito have slightly different requirements. In Australia, *Aëdes amesii* favours rot holes in branches, and *A. vigilax* ephemeral tidal pools in good sunlight to encourage the growth of the phytoplankton on which the larvae feed. Development from egg to adult takes little over a week in the tropics, so a tidal pool left at the height of a spring tide persists long enough for complete larval development. The eggs of *A. alternans* are laid at the edge of pools, and can survive drying out. In East Africa, *Aëdes pembaensis* lays its eggs on the claws of the crab *Neosarmatium meinerti* and the larvae develop within the crab's burrow. Biting midges do not require standing water for breeding; their larvae are part of the substrate infauna. Mosquito larvae flourish in brackish water. Freshwater mosquito larvae have a group of anal papillae which actively acquire sodium and chloride ions from the surrounding medium. Larvae that live in brackish water, such as those found in mangroves, are surrounded by an abundance of sodium chloride, and do not need special mechanisms to acquire it. Their principal physiological adaptation is that the anal papillae are much reduced in size and, in at least one species, they are impermeable to sodium chloride.

Humans are enthusiastically used as prey, but mosquitoes and midges presumably depend largely on the mammal and bird fauna of mangroves. Although there appear to be no records of mosquitoes feeding on reptiles, this is a possibility. Species of *Aëdes* have been seen feeding on mudskippers, so they are not restricted to warm-blooded targets (Hutchings and Saenger 1987; Macnae 1968).

Synchronously flashing fireflies

One of the most unforgettable sights in the mangroves of South-east Asia is the synchronised flashing of fireflies. The River Selangor of western Peninsular

Malaysia is fringed with mangrove trees: as darkness falls, most of these become alive with the lights of thousands of tiny flashes of light. What is most remarkable about these is the regularity of the flashes—about three times a second—and their synchrony. Within one tree all of the flashes appear simultaneously, and neighbouring trees are virtually in synchrony.

In Selangor, the flashes come from the firefly *Pteroptyx tener*. Despite their vernacular name, fireflies are in fact beetles of the family Lampyridae. Of the more than 2000 species of this family found throughout the tropics, only a handful synchronise their flashing. The phenomenon appears to occur only in southern Asia and the western Pacific, from east India through Thailand, Malaysia and Indonesia to the Philippines and Papua New Guinea.

Different species flash with different frequencies. *Pteroptyx malaccae* of Thailand has a cycle of half a second, varying with temperature: at 25°C the timing is 560 milliseconds, with no individuals more than 20 milliseconds out of synchrony (Fig. 3.2). In New Guinea, *P. cribellata* has a cycle of 1 second, and other species flash at intervals of up to 3 seconds.

To achieve synchrony, fireflies must respond to the flashes of their colleagues. Experiments have shown that they can be entrained to the rhythm of a flashing light of appropriate intensity over a range of frequencies, but that the firefly's

Fig. 3.2 Males and females of the firefly *Pteroptyx malaccae* (above). A recording of the sequence of flashes of an individual *P. malaccae* are shown (below, left) and of a group (below, right). a,b: females, c,d: males. The bar on the axis denotes 1 second. (Reproduced by permission from Buck, J. and Buck, E. (1976). Synchronous fireflies. *Scientific American*, **234/5**, 74–85.)

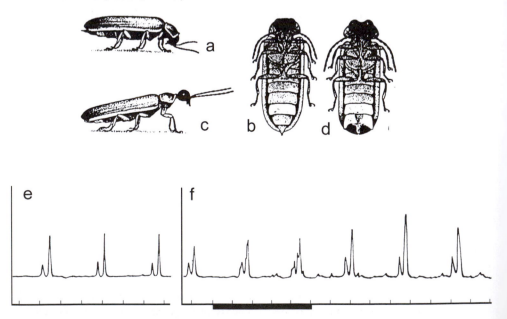

flash coincides with the stimulating flash only at its 'natural' frequencies. A firefly with a natural frequency of one flash per second, stimulated by a lamp flashing at one flash every 1300 milliseconds, will respond by adjusting its frequency to match that of the lamp—but its flashes precede the lamp signal by 300 milliseconds (or lag by 1 second). In natural circumstances, flashing is presumably regulated by an internal pacemaker, whose frequency is character-istic of the species, entrained by the flashes of other fireflies in the vicinity.

Why flash in synchrony? Clearly it must be some form of display. Only certain trees are chosen, mostly *Sonneratia caseolaris*, and (as a subjective impression) those in fairly conspicuous positions. In some areas other mangrove species may be favoured: possibly what makes a tree attractive is the disposition of its foliage, so that a flash can be clearly seen at a distance. Fireflies seem to favour trees which are relatively clear of scale insects and weaver ants: as the two tend to go together it is probably the ants that fireflies are avoiding. A prudent firefly should land in a tree that already has a population of flashing fireflies, advertis-ing that it is ant free: the numbers will therefore accumulate.

Particular trees are consistently used for displays for many years. The displays in one tree in Singapore have been studied for five consecutive years, but reports that Malay fishermen use firefly trees for navigation suggests that individual display trees may be virtually permanent. Individual fireflies probably live for only a few weeks.

Only males flash in synchrony. Although a display tree may have a number of resident males, many more commute in as dusk approaches. Flashing starts soon after sunset, and builds up to a peak within 15 to 20 min. Females approach, with a flickering light: males may enter a brief flash dialogue with females, or even fly off in chase. Copulation may be accompanied by a dull glow by both sexes, then the female may fly off, flashing rhythmically, to seek a site for egg-laying.

It is fairly clear that flashing is used in courtship, but the synchrony of the flashing remains enigmatic. The problem is to work out the individual's advan-tage in participating in group activity. A plausible hypothesis is that flashing establishes a male's credentials as a potentially acceptable suitor, a male that does not correctly synchronise his flashing being excluded from consideration. Once a group of males has assembled round a female, she then selects the one with, for example, the brightest light. This corresponds to the 'lek' courtship of numerous bird and mammal species. The synchrony of whole trees may merely be a consequence of synchronised activity within small groups within sight of each other (Buck and Buck 1976; Macnae 1968).

Other insects

Very little is known of other insect groups within mangrove habitats. It seems likely that the insect fauna of mangrove canopies resembles that of terrestrial forests, and that many species are common to both. Numerous butterfly and moth species have been recorded from mangroves. Bees, including the domestic honey bee *Apis mellifera*, feed heavily on *Aegiceras* and *Avicennia* flowers. Many of these species may be important pollinators (Chapter 1). Other species, such as cockroaches, certainly occur in mangroves, but nothing is known of their ecological importance.

Spiders

Little is known of other mangrove invertebrates of terrestrial origin, other than geographically patchy species lists and more or less anecdotal accounts of some of the more conspicuous species. Among other arthropods, the spiders are best known, or perhaps just more conspicuous. Web-building spiders occur, often in abundance. One of the most spectacular of these is the golden silk spider (*Nephila clavipes*) of the New World. Female *Nephila* build an orb web some 2 m in diameter. The female, which occupies the web, may have a body length of about 6 cm. She may also share her web with a number of kleptoparasitic spiders (*Argyrodes*) which build no web of their own, but steal trapped insects from the *Nephila*.

Another striking group of web-building spiders belong to the family Gastera-canthidae. These have brightly-coloured bodies with projecting spikes which may protect them against predators, and occur in mangroves from the Caribbean to Malaysia and Australia.

As well as the web-building spiders, mangroves often contain numerous wolf and jumping spiders which descend from the trees at low tide and forage over the mud. Their main prey is presumably insects such as ants, but the lycosid *Pardosa* of Malaysia has been seen to catch juvenile fiddler crabs (*Uca*). *Pardosa* is one of the few spiders that seems to be adapted to a semiaquatic life. Its hairy coat is water repellent, and the species seems to shelter (and breed) in air-filled burrows in the mud, including abandoned fiddler crab holes (Stafford-Deitsch 1996).

With this exception (and a few others) the spiders found in mangrove habitats occur also in neighbouring forest. They are not peculiar to mangroves, nor particularly adapted to the mangrove environment.

Vertebrates

Many vertebrate species of terrestrial origin occur within mangrove habitats, particularly reptiles, birds and mammals. Other vertebrates, particularly fish, have entered mangroves from a seaward direction. The provenance of some mangrove species is not straightforward: reptiles such as sea snakes and (perhaps) crocodiles, and mammals such as dolphins are more marine than terrestrial in origin, while some mangrove fish may have originated in fresh water rather than the sea. Nevertheless, for convenience, amphibians, reptiles, birds and mammals are discussed in this chapter, and fish are dealt with in the next among species of marine origin.

Amphibians

Amphibians are rarely found in brackish or salt water. The only exceptions to this are a few species of frog or toad. One of these is the crab-eating frog *Rana cancrivora*, found throughout South-east Asia, where it is sufficiently abundant to be the most important edible frog of the area.

Rana cancrivora is common in rice fields and ditches, often well inland, so is by no means restricted to brackish water. Both adults and larvae are, however, common in mangrove habitats. Rafts of eggs are laid in small pools of water such as crab holes or drainage channels. Young tadpoles, like those of most frog species, are herbivorous: their gut contents include fragments of vegetation, algae and mud. Larger tadpoles have a more mixed diet, and adult frogs are of course exclusively carnivorous. The species lives up to its name: in one study of adult frogs from a mangrove area, every single gut examined contained fragments, or even entire individuals, of the mangrove crab *Sesarma* (see Chapter 4). The diet of *R. cancrivora* from fresh water consists largely of insects (Elliott and Karunakaran 1974).

Given the variety of habitats in which the species occurs, it is clearly euryhaline rather than adapted to a specific salinity range. How does *Rana cancrivora* survive the typical salinities of mangrove habitats?

Tadpoles are effective osmoregulators: that is, they maintain their internal osmotic concentration relatively constant over a wide range of external salinities (Fig. 3.3). This is probably achieved by the active transport of ions out of the body at high environmental salinity, and by the retention of ions and elimination of water when the surrounding salinity is lower than that of the body fluids. Whatever the physiological mechanisms involved, young tadpoles can survive in concentrations of 40 per cent that of sea water (14 parts per thousand, ppt) and older ones to 50 per cent sea water or higher. At 80 per cent sea water (28 ppt) approximately 50 per cent of tadpoles still survive, and some still live at greater ion concentrations than that of full sea water (Gordon and Tucker 1965).

Although the evidence is largely anecdotal, metamorphosis of tadpoles into adult frogs may depend on an influx of fresh water at the start of a rainy season. It has been suggested that tadpoles living in brackish water enter a resting state like the condition of diapause in insects, and that further development, and metamorphosis, are triggered off by a sharp fall in the salinity of the environment (Macnae 1968). This is unlikely to apply in areas such as Malaysia where heavy rainfall is—to say the least—not an unusual or seasonal event.

Adult frogs have a different, and rather more unusual, method of osmotic regulation. Instead of being osmoregulators and maintaining an imbalance between the osmotic concentration of their internal fluids and that of the exterior, they are partial osmoconformers. Internal osmotic concentration is matched with that of the exterior, at least in a hyperosmotic medium. This is brought about not by manipulating ion levels, but by 'topping up' the ionic concentration of the plasma with the non-ionic solute urea $CO(NH_2)_2$ (Fig. 3.4). Other amphibia show slightly raised urea levels in conditions of water shortage, but employing high urea levels to

Fig. 3.3 Total osmotic concentration (continuous line) and concentration of sodium, chloride and potassium ions combined (bars) in blood plasma of tadpoles of *Rana cancrivora* acclimatised to various external salinities. The diagonal straight line represents the line of equivalence between external and internal concentrations. Concentrations are in milliosmoles per litre; figures in brackets indicate sample size. (Reproduced with permission from Gordon, M.S. and Tucker, V.A. (1965). Osmotic regulation in tadpoles of the crab-eating frog (*Rana cancrivora*). *Journal of Experimental Biology*, **42**, 437–45, copyright Company of Biologists Ltd.)

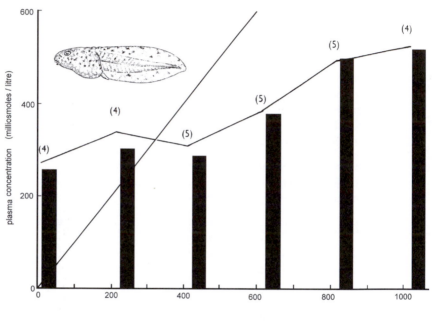

maintain osmotic balance with the environment is a habit shared only with elasmobranch fish (sharks, skates and rays) (Gordon *et al.* 1961).

Urea is the standard nitrogenous excretory product of ordinary adult amphibia, and is normally voided at the earliest convenient opportunity. Indeed, given the solubility of urea, it is not an easy substance for a largely aquatic animal to retain, and it is not understood how *Rana cancrivora* achieves this. Not only is it soluble and difficult to retain, urea is toxic: the concentrations of urea that occur in the frog's plasma at exterior salinities of 80 per cent sea water, should denature enzymes and affect the binding of oxygen by haemoglobin. Somehow *Rana cancrivora* copes with these hazards.

Reptiles

Compared with amphibians, reptiles are relatively common in mangrove habitats. Reptiles habitually occurring in mangroves include numerous species of snake and lizard, and a few species of crocodile.

Fig. 3.4 Total osmotic concentration (continuous line) and concentration of sodium, chloride and potassium ions combined (bars) in blood plasma of adults of *Rana cancrivora* acclimatised to various external salinities. The diagonal straight line represents the line of equivalence between external and internal concentrations. Concentrations are in milliosmoles per litre; figures in brackets indicate sample size. (Reproduced with permission from Gordon, M.S. *et al.* (1961). Osmotic regulation in the crab-eating frog (*Rana cancrivora*). *Journal of Experimental Biology*, **38**, 659–87, copyright Company of Biologists Ltd.)

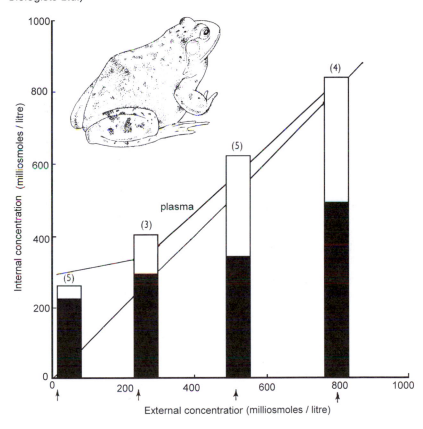

Many snake species are not mangrove specialists, but enter mangroves intermittently from adjacent terrestrial habitats only to forage. Pythons, for instance, may enter Australian mangroves in search of roosting flying foxes (see p. 75). A variety of other terrestrial or arboreal snakes also occur, such as the king cobra (*Ophiophagus hannah*) in South-east Asia, and many lesser species. From the opposite direction, sea snakes (Hydrophidae) also utilise the mangrove habitat. The primitive *Laticauda colubrina* breeds on land, but other sea snakes are fully aquatic, even in their reproduction.

To some snakes, however, mangroves are the primary habitat. Such specialists include the cat snake *Boiga dendrophila* of Australia and South-east Asia, a poisonous and aggressive species that may reach 1.5 m in length. Its dangerous nature is advertised by the classical warning pattern of a glossy black body ringed with gold bands. Mangrove snakes are common in the Indo-Pacific, but also occur in mangrove habitats elsewhere, an example being the Mangrove snake *Nerodia fasciata* of Florida.

Snakes are exclusively carnivorous, and their prey in mangroves consists principally of small fish and crabs. On occasions, foraging snakes have been seen to insert their heads into crab or mudskipper burrows to look for prey. Sometimes the roles are reversed. Judging from analysis of stomach contents, juvenile snakes are an important source of food for crabs. A hard exoskeleton may protect crabs against being bitten, and the ability to respire in water gives a crab the advantage in an underwater dispute over which is the predator and which the prey (Voris and Jeffries 1995).

Many lizards are found in mangroves, usually species typical of adjacent terrestrial forests. In South-east Asia and Australia, monitor lizards are among the more striking: the formidable mangrove monitor lizard (*Varanus indicus*) may reach 1 m in length. As with snakes, the lizards of mangrove habitats are all carnivorous.

Even more formidable than monitor lizards are the crocodiles and alligators. In Central America, the American crocodile (*Crocodylus acutus*) and the common caiman (*Caiman crocodylus*) may occur in mangroves. In West Africa, the Nile crocodile (*Crocodylus niloticus*) is found, while from India through South-east Asia and Australia, as far as Fiji, the commonest mangrove species is the estuarine crocodile (*C. porosus*). Estuarine crocodiles may reach a length of 7 or 8 m. Small crocodiles catch fish and mangrove invertebrates; larger ones also prey on mammals including, should the opportunity arise, humans. The Sarawak Museum recently mounted a display which included, among other memorable items, a wristwatch recovered from the stomach of a crocodile caught in the vicinity a number of years ago.

To live in the mangrove environment, reptiles face the same osmotic problem as other organisms. Figure 3.5 demonstrates the effects of transferring five small American crocodiles between water of different salinities. In sea water there is a loss in mass of 1 to 2 per cent each day as water is lost to the environment. On transfer to fresh water, the lost water is regained and the animals stabilised at around their original mass. At 25 per cent or 50 per cent sea water the animals appeared to be able to maintain their body mass with no net influx or efflux of water (Dunson 1970). Larger crocodiles are better able to cope with sea water than small: one American crocodile thrived after 5 months in full sea water, while the estuarine crocodile has been found in the open sea at some distance from land.

Mangrove-inhabiting reptiles share some common adaptations that enable them to survive an osmotically hostile, saline environment. The keratinised reptile skin is relatively impermeable to water: not completely, but presumably sufficiently so to make it feasible to control water and salt balance by other

Fig. 3.5 Change in mean body weight of five fasting American crocodiles (*Crocodylus acutus*) exposed to different salinities, as indicated. (Reproduced with permission from Dunson (1970). Some aspects of electrolyte and water balance in three estuarine reptiles, the diamond back terrapin, American and 'salt water' crocodiles. *Comparative Biochemistry and Physiology,* **32**, 161–74, © Elsevier Science.)

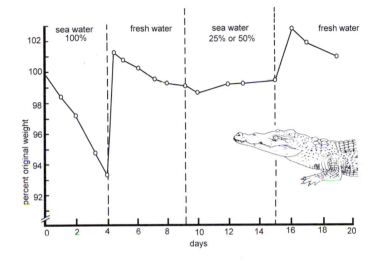

means. Marine snakes are much less permeable to the influx of sodium ions than freshwater species (Dunson 1978). Many species also have valvular nostrils which further limit water movement.

Some water inevitably enters the body with the food. However, in captive conditions feeding takes place primarily when brackish water is available. Drinking is selective. Adult American and estuarine crocodiles avoid drinking when in salt water, and hatchlings may exploit the surface lens of fresh water that results from rainfall (Mazzotti and Dunson 1984; Stafford-Deitsch 1996). Crocodiles are rather more discriminating than alligators in this respect. Given a choice, they will drink only fresh water. Alligators from fresh-water populations are not normally able to discriminate, although animals from estuarine populations of a normally fresh-water species did select fresh water (Jackson *et al.* 1996).

Mangrove snakes show a similar ability to tell salt from fresh water, in contrast to fresh-water snakes which, when placed in sea water, lose water and attempt to compensate by drinking. This makes the problem worse, resulting in catastrophic water loss and death (Dunson 1978).

Even with selective drinking, excess salt inevitably enters the body. The reptilian kidney is unable to excrete urine with a higher salt concentration than that of the blood, although crocodiles (but not alligators) may achieve the same effect by cloacal modification of the composition of their urine (Pidcock *et al.* 1997). Instead of kidneys being the major means of disposing of excess salt, it is eliminated from the body by specialised salt glands. The estuarine crocodile carries between 28 and 40 salt glands, each of which opens by a separate pore on the upper surface of its tongue. Juvenile crocodiles raised in brackish water (20 parts per thousand) have a blood supply to the salt glands three-fold greater than in those reared in fresh water (Franklin and Grigg 1993).

Snakes and lizards also have salt glands. In lizards salty water is secreted into the nasal cavity, in sea snakes and file snakes the glands are sublingual, while in colubrid snakes such as *Cerberus* they are premaxillary in position.

Reptilian adaptations to a saline environment parallel those shown by the mangrove trees themselves. To a large extent salt water is excluded. If fresh water is available, it is selectively imbibed; but if, despite these precautions, it is necessary to eliminate excess salt from the body, then specialised salt glands are employed.

Birds

Birds are highly mobile. Many species spend only part of their time in mangroves, either migrating seasonally, commuting daily or at different states of the tide. They may use mangroves as a feeding area, a nesting area, a refuge from the rising tide or some combination of these. It is accordingly difficult to analyse their significance to the mangrove ecosystem. This difficulty is compounded by the fact that many accounts of mangrove birds are little more than species lists: detailed investigations of the ecology of mangrove birds are sparse.

Waders probe for buried invertebrates on the mud surface, either among the mangrove trees or on adjacent mudflats. At high tide they either leave the area, or perch on mangrove branches or roots until the surface is again exposed. Herons, egrets and kingfishers catch fish in the shallow water of mangrove creeks, or prey on mudskippers (see p. 113, Chapter 4) and crabs on the mud surface. Larger fish-eaters, such as storks, pelicans, ospreys and cormorants, may range further afield and return to the mangroves to roost or breed. The Caroni Swamp of Trinidad is a spectacular example, with densely packed nesting colonies of cattle egrets (*Bubulcus ibis*) and snowy egrets (*Egretta thula*). The scarlet ibis (*Eudocimus ruber*) has been driven by poaching to breed elsewhere, but still roosts in large numbers. The consequence of feeding elsewhere but returning to the mangroves to nest, roost and deposit large amounts of guano is considerable local enrichment with nitrate and phosphate. Trees used as regular roosts by ibises tend to be taller and to have denser foliage (compare p. 21, Chapter 1) (Stafford-Deitsch 1996).

While it is relatively easy to form a general impression of the feeding habits of birds feeding in and around mangroves, it is difficult to quantify their diet, and the relative importance of different items in it. One approach is to catch birds by mist-netting, dose them with an emetic, and attempt to identify the food remains regurgitated. In one study of the birds of mangrove sites in Panama, field observations indicated, unsurprisingly, that three species of egret and six kingfishers foraged on the mud surface in similar ways, and could be regarded as constituting a feeding guild. When individuals were trapped and forced to regurgitate, the diet (based on a total of 46 samples) comprised 45 per cent invertebrate items and 55 per cent fish. The invertebrates included 2.6 per cent gastropod molluscs (snails) and 6.4 per cent crabs: the remainder was made up of spiders, bugs, beetles and various insect eggs and larvae. Clearly these species, generally categorised as fish-eaters, have a fairly varied diet, and continue foraging on mangrove trees and roots when the tide prevents feeding on the mud.

Many mangrove birds, particularly passerines, depend primarily for food on the trees themselves, or their associated invertebrate fauna. Analysis of the feeding behaviour of the birds of Panamanian mangroves, and their regurgitated food, suggested a number of largely insectivorous guilds. Some 33 species, mainly warblers, gleaned insects from leaves; woodpeckers and similar birds—12 species in all—foraged in the bark of the trees; flycatchers (15 species) caught winged insects in the air; and eight species of hummingbird caught insects and spiders by hovering, as well as feeding on nectar (Lefebvre and Poulin 1997).

Analysis of feeding modes in Australian or South-east Asian mangroves indicated similar separation into more or less distinct feeding guilds. One example is shown in Fig. 3.6, describing the foraging behaviour of mangrove birds near Darwin, northern Australia.

Species with similar feeding methods are, actually or potentially, competitors. Within a guild, competition is lessened by some divergence in tactics. In the example shown, for instance, the northern fantail (*Rhipidura rufiventris*) hawked for food in tall, well-spaced *Rhizophora* canopy, while the mangrove fantail (*Rhipidura* sp.) operated in dense, low vegetation at the edges of tidal creeks. The three *Gerygone* species, while very similar in foraging techniques, were segregated by plant species, with the mangrove gerygone (*Gerygone laevigaster*) mainly in low thickets of *Avicennia*, the large-billed gerygone (*G. magnirostris*) favouring the landward *Lumnitzera* zone, and the green-backed gerygone (*G. chloronata*) *Ceriops* and *Lumnitzera* (Noske 1996).

Because guild structure is a function of competitive pressures, and of the pool of species in the area, rather than of intrinsic species-specific characters, the composition of guilds differs from site to site. This is true even comparing sites quite close to each other within Australia. Moving further afield, the mangroves of Peninsular Malaysia and Singapore show a broadly similar pattern of guilds, composed of completely different species (Noske 1995; Sodhi *et al.* 1997).

How similar are mangroves of Central America and those of South-east Asia? Differences in the methods used preclude any simple comparison: for instance, the Malaysian study produced sufficient data on only 17 species, out of a total of about 125 species found in mangroves in the region, whereas in Panama more than 100 species were covered, including many migrants from North America. Despite these major differences, the overall patterns roughly correspond (Table 3.3). In both cases, the greatest number of species forage on leaves, although the number of species was much greater in Panama. Rather fewer species specialised

Fig. 3.6 Dendrogram comparing foraging behaviour of 13 species of mangrove birds near Darwin, northern Australia. (Reproduced with permission from Noske, R.A. (1996). Abundance, zonation and foraging ecology of birds in mangroves of Darwin Harbour, Northern Territory. *Wildlife Research*, **23**, 443–474, © CSIRO.)

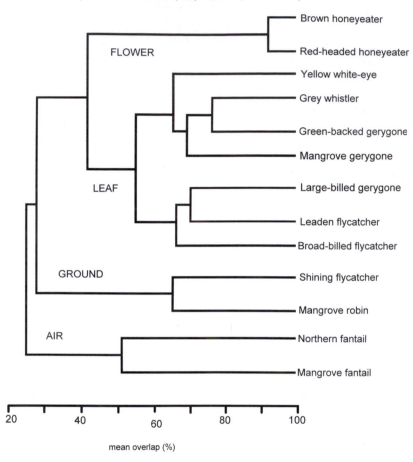

mean overlap (%)

Table 3.3 Distribution of insectivorous bird species between the four major foraging modes, data from Noske (1995) and Lefebvre and Poulin (1997)

	Malaysia	Panama
leaves	7	33
flowers	5	8
bark	4	12
aerial	1	15

in foraging on bark. Among the flower specialists, the sunbirds of South-east Asia are counterparts of the hummingbirds of the Americas, and feed similarly on nectar and insects. Relatively more species in Panama than in Malaysia feed on aerial insects.

To most species of birds found in mangroves, the habitat is a convenient feeding resource, but not the only one on which they depend. Long-distance migrants, such as many of the warbler species found in mangroves of South-east Asia and Central America, and the millions of birds of all sorts that annually migrate from central Asia to the Indus delta (see p. 193, Chapter 8), spend only part of their year in mangroves, and the remainder in very different habitats.

Other species may move shorter distances to occupy mangroves on a seasonal basis. In Panama, flooded habitats such as mangroves are buffered from the effects of seasonal rainfall and the resulting variation in availability of insects. Forest-living insectivorous birds therefore move into mangroves when rainfall, and food, are sparse in alternative habitats. In Malaysia, sunbirds (Nectariniidae) move into mangroves when the trees are in flower. As their family name suggests, they feed on nectar, as well as on small insects. When flowering ceases, the birds move out into adjacent rainforest. Sunbirds are the principal pollinators of several species of *Bruguiera*. In Australia they are replaced by the similar honeyeaters (Meliphagidae), and in Central America by hummingbirds (Trochilidae) (Noske 1995).

A few species are more or less totally dependent on mangroves, and are rarely found elsewhere. In Australia, for instance, more than 200 species of bird have been recorded from mangroves. Of these, only 14 species of passerine are virtually confined to mangroves, with a further 11 species found predominantly in mangroves, but frequently occurring elsewhere. A similar pattern obtains elsewhere: in New Guinea with 10 species, and Malaysia, with around eight species regarded as mangrove specialists. Even these figures may be misleading: of the apparent 'mangrove specialists' in Malaysia, four occupy different habitats in other parts of the world, one example being the cosmopolitan great tit (*Parus major*) which is distributed from western Europe to China. Elsewhere in the world, even fewer species are restricted to mangroves: in Africa, a single species of sunbird, and in Trinidad and Surinam a single species of crab-eating hawk (Noske 1996).

Why do so few birds specialise in the mangrove habitat? Africa and Central America have fewer species of mangrove tree and, smaller areas of mangrove, than Australia, or South-east Asia. Either factor might contribute to the paucity of mangrove specialists. However, even in the extensive mangroves of South-east Asia and New Guinea relatively few of the bird species found in mangroves are mangrove specialists. Why is this?

One reason might be that there is little scope for niche specialisation, compared with terrestrial forests. The physical structure of a mangrove environment is simpler than that of most tropical rainforests, with virtually no understorey, less layering of the canopy, and a degree of convergence in leaf size and shape between the different tree species (p. 17, Chapter 1). In addition, regular inundation of the forest floor eliminates the possibility of ground-living specialisation.

Adaptation, as generalists, to a relatively simple forest structure may be the principal reason why mangrove birds seem to be better than rainforest ones at colonising man-made habitats. Villages, plantations, urban parks and gardens share with mangroves a relatively open structure with few vegetation layers and relatively few tree species (Noske 1995).

A second possibility is that mangroves are typically unstable habitats, characterised by accretion and erosion, and the rapid colonisation of canopy gaps (p. 44, Chapter 2). Mangrove birds may tend, as a result, to be adaptable and capable of rapidly exploiting new opportunities, rather than narrowly specialist and restricted in their requirements.

Finally, the interchange of species between mangroves and other habitats, such as rainforest, may have limited the emergence of mangrove specialists. It is noticeable that there are proportionately fewer mangrove specialists in New Guinea, where rainforest usually abuts mangrove forest, than in Australia, where this is less common. Within Australia, there are fewer specialists in the mangroves of Queensland, which are extensive and contiguous with rain forest, than in north western Australia where this is not the case. Mangrove specialists are more likely to arise where there is geographical isolation to prevent gene flow between an incipient specialist and its parent species, and where there is less exposure to rainforest competitors or invading migrants. Conversely, some species may have become restricted to mangroves by being squeezed out of rainforest by more effective competitors (Ford 1982).

Looking at the ecology of the mangrove ecosystem as a whole, it makes little difference whether a particular resident bird species occurs also in other habitats: abundance, diet and total food intake are more important considerations. Although there is little reliable data on abundance, population densities of mangrove birds appear in general low compared with, say, rainforest. This may be related to the relatively simple forest structure, tidal restriction of feeding time, and possibly limited nesting sites. Migrant and commuter species have more unpredictable effects. A species that feeds elsewhere, and resorts to mangroves to roost or breed, is likely to provide a net input of nitrates and phosphates. In contrast, flocks of seasonal migrants which use mangroves to replenish energy reserves for a long passage, may represent a significant export of carbon. Even less is known—much less quantified—about the role of transient bird populations than about that of residents.

Mammals

As with birds, few mammals are characteristic of mangrove habitats. Most species found in mangroves are those that occur in adjacent habitats, although in some cases mangroves harbour species that have been eliminated elsewhere. Those that do occur in mangroves avoid immersion by being either highly mobile or tree-living.

The only completely aquatic mammals found in mangrove creeks and rivers are dugongs and cetaceans. Most coastal mangroves are periodically visited by dolphins and porpoises. In the great Sundarban mangroves, lying between India and Bangladesh, there are also the rare freshwater Ganges and Irrawaddy dolphins (*Platanista gangetica* and *Orcaella brevirostris*).

Otters, also fish eaters, are common in South-east Asian mangroves. Other small carnivores may visit mangroves in search of fish or crabs. These include the fish cat (*Felis viverrima*) and mongooses (*Herpestes* spp.) in Asia, bandicoots (*Perameles* spp. and *Isoodon* spp.) in Australia and the raccoon (*Procyon cancrivorus*) in Central America.

Terrestrial herbivores forage for mangrove seeds or foliage at low tide. Examples in South American mangroves are the agouti (*Dasyprocta*) and the key deer (*Odocoileus*). In other parts of the world antelopes, deer, wild pigs, and rodents may be common. In the Sundarbans pigs and the axis deer (*Axis axis*) constitute the main prey of the Bengal tiger (*Panthera tigris*): until relatively recently the Javan rhinoceros (*Rhinoceros sundaicus*), water buffalo (*Bubalus*) and swamp deer (*C. duvauceli*) were also common. Domestic camels and buffalo are major eaters of foliage in the mangroves of Arabia (Fig. 3.7) and in the Indus delta of Pakistan (Chapter 8).

Fig. 3.7 Camels feeding on mangrove foliage in the Arabian Gulf. (Photo A.R. Dawson Shepherd.)

None of these species is a mangrove specialist. Their association with mangroves is either temporary, or—as with the Bengal tiger—a result of their elimination by man from alternative terrestrial habitats. The species that is perhaps closest to being a true mangrove mammal is the rare Australian rodent *Xeromys myoides*. This small rat forages for crabs at low tide among *Avicennia* and *Rhizophora*, and builds a nest of leaves and mud among the buttress roots of *Bruguiera*, above the level of neap high tides. Although its fur is water repellent, it has not been seen to swim, so probably avoids high tides by climbing trees (Hutchings and Saenger 1987).

Monkeys are often common in mangroves. Some are omnivores: in southern Senegal (west Africa) the vervet monkey (*Cercopithecus*) eats both fiddler crabs (*Uca tangeri*) and *Rhizophora* flowers, fruit and young leaves, while in South-east Asia macaques (*Macaca*) forage on the mud for crabs and bivalve molluscs. They are also in the habit of uprooting *Rhizophora* propagules. Few of these are even slightly damaged, much less eaten, so the motivation is not clear. Whatever the reason, monkey damage is a major problem for mangrove replanting projects (see Chapter 8), and may also be a significant factor in natural regeneration and community structure of mangrove ecosystems.

Some monkey species are more exclusively herbivorous. In South-east Asia, these include colobine monkeys such as langurs or leaf monkeys (*Presbytis*) and the striking proboscis monkey (*Nasalis larvatus*). Although their diet includes fruit, colobines are specialised in eating leaves. They have a specialised stomach divided (like that of ruminants) into a number of chambers. The forechambers contain anaerobic bacteria, capable of digesting cellulose and dealing with the antifeeding substances produced by plants, which deter many herbivores. Mangrove leaves are particularly rich in tannins and similar feeding deterrents (p. 87, Chapter 4). *Sonneratia* appears to be the preferred mangrove species, where available, followed by *Avicennia* (Salter *et al.* 1985).

Proboscis monkeys are exclusively found in Borneo and are restricted to mangroves and riverine forests. They form small social groups, typically with a single male. The male has the prominent pendulous nose—which shoots out into a horizontal position when the male honks to deter intruders—that gives the species its common, and scientific, names. It has been estimated (in an inland riverine forest, rather than a mangrove area) that the average density of proboscis monkey groups is 5.2 km^2. Taking into account the mean body mass of the monkeys, this works out at an average monkey biomass per square kilometre of just under 500 kg. Experiments on captive monkeys suggest a daily food intake of around 12 per cent of body weight, explaining their pot-bellied appearance and almost continual feeding. This amounts to 60 kg of leaves and fruit per square kilometre each day, or more than 21 t annually—an impressive total, and an indication of the potential ecological significance of the species (Dierenfeld *et al.* 1992; Yeager 1989).

The abundance of insects in mangroves attracts large numbers of insectivorous bats (Microchiroptera), and a mangrove forest may be exploited by numerous species. Competition results in species foraging in different microhabitats within the mangal. Table 3.4, for instance, shows how the aerial environment was partitioned between the bat species of the mangroves of Kimberley, Western Australia. Even species apparently foraging in the same zone showed differences in technique: when measures of wing morphology—hence of flight pattern—were taken into account, there was virtually no overlap between species (McKenzie and Rolfe 1986). The impact of insectivorous bats on the insect population must be considerable. Judging from estimates for temperate species, a single bat may consume between one-quarter and one-third of its body weight in insects in a single night. A bat weighing 30 g might therefore eat 10 g of insects, amounting to perhaps 5000 individuals, each night.

Fruit bats and flying foxes (Megachiroptera), exclusively Old World, may also be abundant in mangroves, using the trees as a roost as well as a source of food. Roosts of flying foxes (*Pteropus*) in Australia have been estimated to comprise as many as 220 000 individuals (Macnae 1968). The location, as well as the vast numbers, protect against many potential predators.

Most fruit bats feed on nectar as well as on fruit, and it is this which attracts many species to feed in mangroves. In Peninsular Malaysia, two of the more important species are the long-tongued fruit bat *Macroglossus minimus*, and the cave fruit bat *Eonycteris spelaea*, which eat *Sonneratia* nectar and pollen. Long projecting stamens deposit large amounts of pollen on the fur of the bat: the species is a major pollinator of *Sonneratia*. Possibly because of the wear and tear resulting from each visit by a bat, the flowers of *Sonneratia* last for only a single night, and each morning the mud beneath a *Sonneratia* stand is covered by shed stamens.

Macroglossus has a very close association with *Sonneratia*, and in western Malaysia has never been recorded away from mangrove areas. Specialisation in *Sonneratia* is possible because, at least in one Malaysian site studied, *Sonneratia caseolaris* is available as a food source throughout the year, while two other species of

Table 3.4 Time allocation of insectivorous bats foraging in Kimberley mangroves, based on position of bat when observed or captured, from McKenzie and Rolfe (1986)

	OC	AC	BS/O	BS/A	IS	N
Taphozous flaviventris	88	10	2			42
T. georgianus	90	10				20
Chaerephon jobensis	82	16	2			44
Mormopterus loriae	14	55	32			22
Pipistrellus tenuis			16	79	8	36
Chalinolobus gouldii		8	69	15	8	13
Nycticeius greyi		2	24	52	22	46
Nyctophilus arnhemensis				14	86	51

The figures are percentages of the total observations (N) made on each species. OC: more than 4 m above canopy, AC: just above (< 4 m) canopy, BS/O: beside stand in the open, BS/A: beside stand but against surfaces, IS: inside stand.

Sonneratia (*S. alba* and *S. ovata*) flower in flushes and collectively were available for about three-quarters of the year.

Eonycteris has a different strategy. It roosts well inland. One massive *Eonycteris* roost is in the limestone caves of Batu, not far from Kuala Lumpur. Tens of thousands of bats may be present. Analysis of pollen in guano from these caves showed that *Sonneratia* was one of the most frequented species. Batu is 38 km, as the bat flies, from the nearest mangroves, so the nightly foraging range is considerable.

Many other species of plant are visited. One of these is the commercially important durian (*Durio zibethinus*), a fruit described by some as delectable, and by others as redolent of sewage. *Eonycteris* is a major pollinator. Durians flower and fruit for only limited periods, and pollinating bats therefore require alternative sources of food. Were it not for the availability of *Sonneratia* as a stopgap, durians would not be pollinated and the crop would fail (Start and Marshall 1976).

Both insectivorous and nectar-eating bats cover large distances when feeding, and neither depends exclusively on mangroves. For this reason, they represent important links between mangroves and terrestrial habitats: the mangrove habitat cannot be viewed in isolation from its surroundings.

4 The mangrove community: marine components

Mangroves offer the same opportunities to organisms of marine origin as to those of terrestrial origin discussed in the last chapter, providing both a physical environment and a source of nutrients. Pneumatophores and prop roots greatly expand the surface area available, and provide a hard substrate, in contrast to the surrounding mud, while the primary production of mangroves supplies an energy source to many organisms.

Algae

Apart from the hard substrate offered by mangrove pneumatophores, aerial roots and trunks, the mangrove soil also provides a surface on which photosynthetic algae can grow. For the most part, these are unicellular diatoms. Bluegreen Cyanobacteria (often erroneously called bluegreen algae: algae being eukaryotes, while Cyanobacteria are prokaryotes), are also present on almost any available surface. These, too, are photosynthetic. In addition to their role as primary producers, microscopic algae alter the texture of the soil by affecting particle size and binding soil particles together with their mucous secretions, while some bluegreens fix nitrogen from the atmosphere (see also Chapter 1).

Macroalgae are rarer on the mud surface, as most species require a firm surface on which to attach. There may, however, be permanent patches of unattached red algae (Rhodophyta) *Gracilaria* and *Hormosira*. *Gracilaria* is extensively cultivated for the extraction of agar, and the natural populations found in some Malaysian mangroves are so abundant that their commercial exploitation has been considered. *Hormosira* is sometimes so dense on mudflats adjacent to mangroves that it reduces the chances of successful settlement by *Avicennia* seedlings by as much as by 80 per cent, so inhibiting the spread of mangroves into otherwise potentially suitable areas (Clarke and Myerscough 1993).

Pneumatophores and aerial roots are colonised by a film of diatoms and unicellular algae, as well as by a turf of small red algae. Most characteristic is a community of *Bostrychia*, *Caloglossa*, *Murrayella* and *Catanella* which, in various permutations of species, is found virtually throughout the tropics. The assemblage is often termed a bostrychietum, after its principal constituent. These species seem tolerant of a wide range of salinity conditions and *Bostrychia*, at least, is highly resistant to desiccation. All three genera grow better in shady conditions. In sunlit patches the dominant species are green algae (Chlorophyta) such as *Enteromorpha*.

An accumulation of algae probably impedes the efficiency of aerial roots and pneumatophores by blocking lenticels, although this has not been directly demonstrated. Certainly algal growth decreases the survival rates of *Rhizophora* seedlings, and algae must be periodically removed to ensure the success of replanting programmes (Chapter 8).

Fauna of mangrove roots

Mangrove roots also host a range of animal epibionts, in their subtidal and tidal parts. The distribution of these often shows a pattern of zonation like that familiar from studies of the sessile fauna of other hard-bottomed shores. Barnacles (*Balanus* spp.) are among the most conspicuous fouling organisms on the aerial roots of *Rhizophora* and the pneumatophores of *Avicennia*. They may even settle on leaves rather than on the trunk in situations where these are submerged at higher tides: a short-sighted policy, given the relatively rapid turnover of leaves compared with the trunk surface. Barnacles are filter feeders, hence they depend directly on mangroves only for a stable anchorage.

In some circumstances, however, encrusting barnacles can significantly affect root growth. Aerial roots seem to accumulate rather few barnacles. When an aerial root reaches the mud surface and spreads laterally, however, barnacle cover is more likely to be high enough to block lenticels and reduce gas exchange and respiration. In one study of the epibiont community on the roots of the red mangrove (*Rhizophora mangle*) on the Pacific coast of Costa Rica, barnacle coverage of 75 per cent was shown to reduce root growth rate by 30 per cent, and net production by 52 per cent.

The density of barnacles on mangrove roots is high only when their normal predators are absent. In the study just described, this was achieved experimentally by excluding them with mesh cages. The most important barnacle predators in the area are the snails *Morula lugubris* and *Thais kiosquiformis* and the hermit crab *Clibanarius panamensis*, and when they are absent barnacles flourish and root growth is inhibited. Interaction between a filter-feeding epibiont and its predators therefore materially affects *Rhizophora's* ability to produce roots.

As well as encrusting organisms such as barnacles on the outside, mangrove roots are also afflicted with animals which burrow into the root tissue. These include, in particular, molluscs of the families Teredinidae (shipworms) and the Pholadidae (piddocks), and the isopod crustacean *Sphaeroma*. Teredinids are particularly important in destroying dead wood (p. 109). As with many other herbivorous animals, they cannot themselves digest wood. Like termites, they contain symbiotic cellulose-digesting and nitrogen-fixing bacteria which carry out the difficult biochemistry of dealing with the more recalcitrant plant materials.

Wood-borers can be highly destructive when they feed on the timbers of wooden harbour installations or boats, and are no less so when they attack the living tissues of mangrove trees. The aerial portions of roots are particularly badly affected by *Sphaeroma*, probably because closer to the mud surface predators protect roots by preventing the isopod larvae from settling. The effect on aerial roots can by significant. In one study, isopod infection reduced the growth rate by half, and net root production by 62 per cent (Perry 1988).

On the other hand, *Sphaeroma* can potentially benefit mangroves. Damage to the growing root tip can result in branching, so that the number of roots actually reaching the soil surface is increased. Enhanced anchorage will make a tree more stable, possibly an important consideration in an area affected by hurricanes (Simberloff *et al.* 1978).

The subtidal portions of *Rhizophora* roots usually carry a much richer growth of sessile epibionts, animal and plant. As well as barnacles, these can include oysters (*Crassostrea*) and other bivalve molluscs, sponges, tunicates, serpulid and sabellid annelid worms, hydroids and bryozoans. These in turn attract a variety of mobile browsing or predatory animals. There is no evidence of adverse effects of subtidal fouling organisms on their host: indeed, species of encrusting sponge have been shown to benefit their host by supplying it with soluble nitrogenous nutrients (Ellison *et al.* 1996; see p. 19, Chapter 1). Possibly other such interactions remain to be discovered.

If mangrove root fouling communities are considered in terms of functional groups, they appear relatively consistent and predictable. The dominant organisms are those which filter feed close to the surface, such as barnacles, ascidians, bryozoa, tube-building polychaete worms and bivalve molluscs. There are a number of predators, such as thaidid molluscs, and some more general foragers, including species of crab. These broad categories, however, contain many species and a great deal of variation in space and time. The richness and diversity of the assemblages of mangrove root fauna, and the considerable variation between sites, and even between neighbouring roots, raise questions of what factors determine community structure and the species present.

Broad biogeographical considerations have an influence. The mangrove root communities of the tropical western Atlantic and Caribbean are dominated by sponges and tunicates, while those of similar latitudes in the eastern Pacific have abundant barnacles and low coverage of sponges and tunicates. In the Indian Ocean and west Pacific, oysters are often more abundant. In total species numbers, the Caribbean communities are apparently richest, with more than 100 species of invertebrate and algal epibionts recorded. This is surprising, since the Indo-West Pacific, as a biogeographical region, is in general more species-rich. The conclusion could, of course, be misleading and is perhaps due simply to greater effort put into investigating root communities in the Caribbean than elsewhere.

The processes that structure a community are complex and varied, and different species respond differently, and at different scales of time and space. Species distributions are affected by gradients in physical variables: either at the vast scales of climate and latitude, the vertical scales of a tidal range, or at even smaller scales. Communities may show successional changes, culminating in a relatively stable and predictable community as rapid colonisers are replaced progressively by more effective competitors. Periodic disturbances, such as storms, may frustrate this process and maintain a balance between colonisers and competitors. Smaller-scale and probably more frequent physical disturbances, or the action of grazers or predators, can create gaps, filled sporadically by whatever colonising species are available. Selective predation, of course, can alter the balance between prey species. Or the composition of a community can be dominated by chance and the vagaries of the supply of larvae in the plankton: most encrusting marine animals have a planktonic larval stage. Those that do not, or whose larvae disperse only short distances, can dominate areas for no reason other than proximity of a parent.

Which of these general considerations are most important to mangrove root-fouling communities? The most illuminating analyses of community structure come from studies of mangroves in the western Atlantic region, particularly in the Belize cays. Variation in community structure on *Rhizophora* roots was investigated over spatial scales ranging from around 1 cm (between front and back of the same root) to 10 km (comparisons between cays), and over a period of more than a year. The results are complex, but the general conclusion is that the variable supply of larvae is important in determining the structure of the epibiont community over short time scales, and over small and very large spatial scales. Over long time scales, and at intermediate spatial scales, variation in physical factors may be more important (Farnsworth and Ellison 1996).

What applies to the mangrove root communities of the Belize cays may not, of course, apply in other areas. In the Florida Keys, mangrove root fouling communities vary a great deal. Much of the variation is probably due to physical disturbance, particularly from strong tidal flows (Bingham and Young 1995). In contrast, root epifaunal communities on mangroves in Venezuela (again, *Rhizophora*) are highly stable in abundance and species composition. This is possibly because, unlike Florida, the tropics are virtually aseasonal, and factors such as the supply of colonising larvae therefore do not fluctuate greatly (Sutherland 1980).

Invertebrates

Animals of the root-fouling community use mangroves primarily as a hard substrate; a suitable site for attachment. Many invertebrates depend on mangrove productivity either directly or indirectly. Some species eat shed leaves or reproductive products. Others ingest fine organic particles, either suspended in the water column, as filter feeders, or from the settled mud. Others again are predators or general scavengers. Mangrove invertebrates include representatives of many phyla: molluscs, arthropods, sipunculans and nematode, nemertean, platyhelminth and annelid worms. The most abundant and conspicuous, however, are crustaceans and molluscs.

Crustacea

Among the most abundant and diverse of mangrove crustacea are the Brachyura, or true crabs, and among the brachyura the dominant mangrove species in many parts of the world belong to two related families, Grapsidae and Ocypodidae. The latter includes the fiddler crabs (*Uca* species). Other crustacea commonly associated with mangroves include a few predatory swimming crabs of the family Portunidae, hermit crabs, the commercially important penaeid shrimps (Chapters 6 and 8), a variety of burrowing species such as the mud lobster *Thalassina*, amphipods and isopods.

Crabs

The family Grapsidae contains numerous species world-wide. Many show a tendency to live in brackish, or even completely fresh water; some are amphibious and a few virtually terrestrial. It is not surprising, therefore, to find that many grapsids are characteristic of mangrove habitats where they forage at low tide on the exposed mud, or on mangrove pneumatophores or tree trunks. At high tide they avoid predators such as fish by retreating into burrows in the mud. Some are more active by day, others by night.

Among the grapsid crabs, the most important, in number of species and abundance, belong to the subfamily Sesarminae, particularly to the genus *Sesarma*. These are found throughout the tropical and subtropical areas of the Atlantic, Indian and Pacific Oceans, particularly in mangrove habitats. Densities of 50 to 70 per m^2 (equivalent to a biomass of up to 20 g m^2) are not unusual. Other grapsid crabs are also characteristic of mangroves (e.g. *Goniopsis* species in the Caribbean), while some such as *Metopograpsus* occur frequently in mangroves but are also common in other habitats (Fig. 4.1).

Sesarma species (using *Sesarma* in the broadest sense) closely resemble each other. The most distinctive features are a trapezoidal-shaped carapace, not heavily calcified, typically showing a pattern of fine lines of short setae. Ventrally, the carapace plates that lie towards the front of the body, below the orbits, have a well-marked, regular, grid-like pattern of short bristles. This structure is important in respiration and, probably, in temperature control (see p. 100). The carapace is often cryptic in colour, matching the colour of the mud, although the chelae may be brightly coloured, presumably reflecting some form of signalling function. When disturbed, sesarmines generally show considerably more agility than biologists. Because of this, and because they (and the biologist attempting to catch them) are frequently covered in mud, they are not the most amenable of animals to identify in the field. Some representative sesarmines are shown in Fig. 4.2.

Fig. 4.1 A grapsid crab (*Metopograpsus* sp.) foraging on an *Avicennia* pneumatophore. (Photograph HAR.)

Fig. 4.2 Species of sesarmine crab: (a) *Parasesarma plicata*, (b) *Aratus pisonii*, (c) *Neosarmatium smithi*. *Neosarmatium* shows the grid-like structure of the epimeral plates, involved in the recirculation and oxygenation of water from the gill chambers. The bars indicate 1 cm. (From Jones, D.A. (1984). Crabs of the mangal ecosystem. In *Hydrobiology of the mangal* (ed. Por, F.D. and Dor, I.), pp. 89–109, Fig. 1 (part), © 1984, Dr W. Junk publishers, The Hague, with kind permission from Kluwer Academic Publishers.)

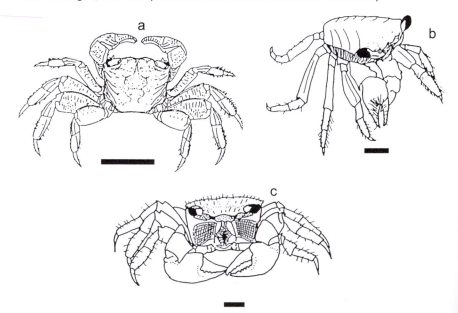

Because of the similarity in morphology, sesarmine taxonomy is—to say the least—complex and confusing, particularly in relation to the species of South-east Asia and Australia. Many species are inadequately described, or have never been fully described. Of 37 species of sesarmine recorded from Australia, for example, three were undescribed species and a further nine could not definitely be assigned to a known species. Sadly, this situation is fairly typical, and one that mangrove biologists learn to live with. Overall, probably about 20 per cent of mangrove crab species in Australia are unnamed (Davie 1982). Some sesarmine species have been described more than once, independently, under different names, and the genus has been split into subgenera, not all of which are universally accepted. In some instances subgenera have been promoted to full genus status by some authors, but not by others. Progress is being made towards rationalising this taxonomic nightmare, but the position is still far from clear. The situation is slightly better in other parts of the world. I shall take a fairly robust view of the taxonomic niceties and generally avoid repetition by using subgeneric names.

Leaf eating by crabs

Grapsid crabs are often fairly general feeders, depending mainly on scavenging and a certain amount of predation. The family as a whole, however, shows a distinct propensity for herbivory. This is particularly true of *Sesarma* and its relatives. Herbivory in sesarmines may include the scraping of epiphytic algae from the surface of mangrove roots, trunks or branches. The most important form of crab herbivory, however, is the eating of leaves and the reproductive products of mangrove trees. It is of little importance to the tree what happens to fallen leaves, and to the discarded components of flowers. Removal of live and photosynthetically active leaves does happen, but is relatively rare, while the destruction of propagules is frequent and highly significant.

Of the various contributions mangrove trees inadvertently make to herbivorous crabs, fallen leaves are probably quantitatively the greatest. As a result, leaf eating has been extensively studied. The most important herbivorous crabs are undoubtedly sesarmines, but other species must also be considered. The Indian Ocean land crabs *Cardisoma* and *Gecarcoidea* (family Gecarcinidae), and the Central American hairy land crab *Ucides* (family Ocypodidae) are locally important eaters of mangrove material.

Crabs may manoeuvre entire leaves into their burrow, or eat leaves or leaf debris on the mud surface. Food is ingested by manipulating and positioning it with the claws and external mouth parts. It is then broken into smaller portions with the mandibles and passed into the stomach. A crab's stomach contains an array of ridges and hooked teeth, which act against each other and grind the food into smaller particles, which are then passed back down the gut for digestion and absorption (Fig. 4.3). Some crabs secrete enzymes such as cellulase: this has not been demonstrated in *Sesarma*, but would certainly be useful in breaking down plant material.

Fig. 4.3 Gastric teeth of (a) the herbivorous sesarmine *Aratus pisonii*, (b) a deposit-feeding fiddler crab (*Uca* sp.) and (c), for comparison, a carnivorous portunid swimming crab *Callinectes sapidus*. The dorsal and left lateral tooth are shown in each case. (From Warner 1977).

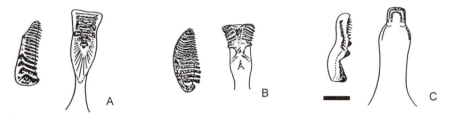

Analysis of stomach contents shows the importance of mangrove leaves to the diet. In many cases the stomach (strictly speaking, the foregut) seems to be stuffed full of leaf fragments. In terms of volume, very few other constituents of the diet appear significant—although, of course, an item minor in volume may be of major significance if it supplies a requirement (such as nitrogen) in which plant material is deficient. Typical estimates in some representative species indicate that more than 90 per cent of the foregut contents may consist of mangrove leaf fragments (Table 4.1). The diet of sesarmines is dominated by mangrove leaves.

Direct measurement of feeding rates shows that South African *Neosarmatium meinerti*—a relatively large species—can eat up to 0.5 g dry weight per day of mangrove leaf material, equivalent to an energy intake of approximately 10.0 kJ d^{-1}. Taking into account the weight of the crab, typical feeding rates are in the region of 15 to 36 mg dry weight of leaf per g crab per day (Emmerson and McGwynne 1992).

Analysis of gut contents can be misleading. Ingestion does not necessarily mean that mangrove material is actually assimilated: it might pass through the gut without being absorbed. Assimilation efficiencies of mangrove leaves by sesarmine crabs have been tested by feeding experiments in the laboratory, and are not high. One study suggested figures ranging from 9 to 32 per cent, depending on the age and state of the leaves, leaves that had spent some time on the mud

Table 4.1 Principal components of the diet of representative sesarmine crabs, given as estimated percentage by volume, of the foregut contents, data from Leh and Sasekumar (1985) and author

	Higher plant	Other plant	Animal	Sand, mineral	Other, unidentified
Sesarma eumolpe	91	1	4	4	0
S. onychophora	83	1	13	2	1
S. plicata	67	1	1	5	27
Episesarma versicolor	90	0	7	3	0
Sesarmoides kraussi	93	0	5	2	0
Cleistocoeloma merguiensis	97	0	2	1	0

being more readily assimilated than those removed directly from trees (Kwok and Lee 1995).

Extrapolating from laboratory experiments is risky. The crabs had little choice but to eat the leaves offered to them, and the absorption efficiencies measured might not reflect their true extent of dependence on mangrove material in natural circumstances. There is now one method which avoids the difficulties of interpreting gut contents and artificial feeding experiments, and theoretically enables an assessment of the overall importance of mangrove material in the diet of sesarmine crabs. This depends on the relative proportions of different stable isotopes of carbon in mangrove leaves and in the tissues of organisms thought to depend on them.

The carbon fixed by mangrove leaves during photosynthesis is, of course, predominantly the most abundant isotope, carbon-12 (^{12}C). A small proportion is ^{13}C: the exact proportion varies between mangrove species, but mangroves as a group have a characteristic ^{13}C/^{12}C ratio in their tissues which differs from the ratio found in, for example, seagrasses or algae. The ratio, relative to a standard, is expressed as a δ^{13}C value, in parts per thousand. Mangrove leaves have δ^{13}C values in the range -24‰ to -30‰. When organic matter is assimilated by an animal, there is some preferential absorption of ^{13}C rather than ^{12}C, but this effect is small and raises the δ^{13}C value (makes it less negative) by perhaps only 1 to 2‰ at each trophic level. This means that the δ^{13}C value of an animal's tissues is a good indicator of the value in its food.

Provided there are only two alternative food sources, and that their δ^{13}C signatures are known, and differ significantly from each other, δ^{13}C analysis can be used to ascertain which is the dominant food source, and even to estimate its relative contribution. By comparing δ^{13}C values of different organisms, it is even possible in some situations to track mangrove carbon through successive stages of food chains.

Figure 4.4 shows the distribution of δ^{13}C values of samples of mangrove and algal material, and in the tissues of various other mangrove animals. Only in gastropod molluscs and a few grapsid crabs (mostly sesarmines) do the δ^{13}C values overlap with those of mangroves, confirming the importance of mangrove material in the diet of these species (Rodelli *et al.* 1984).

Are crabs selective feeders?

In most mangrove areas, there is a 'standing crop' of leaf litter, of varying age and state of decomposition, available to a foraging crab. By analogy with the term 'biomass', which refers to the mass of living organisms, this is sometimes referred to as 'necromass'. Necromass is made up of fallen leaves varying in size, physical toughness, nutritional value and palatability, depending on the species of tree from which they come, and the time for which they have been decomposing after their fall. Do crabs feed selectively, or indiscriminately, from this varied menu?

Fig. 4.4 Distribution of $\delta^{13}C$ values of samples of mangrove and algal material, and in the tissues of animals collected from within mangrove swamps, coastal inlets and offshore. (From data in Rodelli *et al.* (1984). Stable isotope ratio as a tracer of mangrove carbon in Malaysian ecosystems. *Oecologia*, **61**, 326–33, © Springer-Verlag, reproduced with permission.)

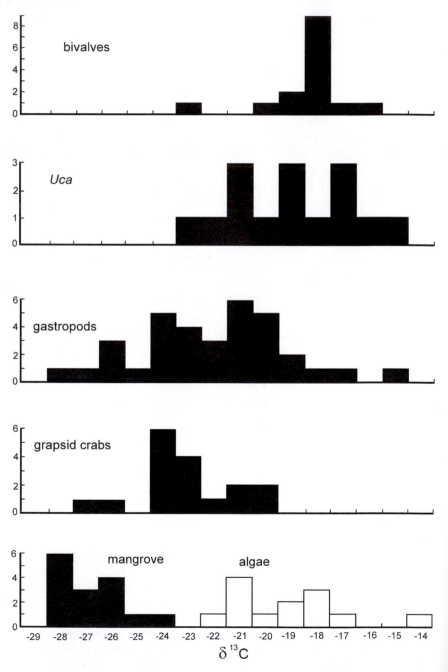

Preferences have been established in the crab *Neosarmatium smithi*, a large sesarmine abundant in the mangroves of Queensland. Freshly picked and senescent leaves from *Ceriops tagal* trees were allowed to age on the mud surface for different periods of time, before being presented to crabs, in different combinations. After 24 h, the leaf fragments that remained were sorted into their original categories and the amount of each category eaten was worked out by subtraction from the amount present at the start of the experiment. When the crabs' choice was between fresh, senescent and decayed leaves, they clearly preferred decayed (Table 4.2).

Why are decayed leaves preferred to fresh or senescent ones? A further series of experiments suggest an answer to this question. In this case, crabs were fed each leaf type separately—newly-picked, and decayed for various periods of up to 10 weeks—and the results expressed as the amount eaten relative to the total for all diets. Figure 4.5 shows the increasing relative consumption rate as the leaves progressively decay.

The same investigators also recorded the changing composition of decaying leaves. As decay proceeded, the leaf content of flavolans decreased. Flavolans, or condensed tannins, are present in relatively high concentrations in mangrove leaves, for the most part covalently linked to glycan carbohydrates as flavoglycans. They make leaves unpalatable or indigestible. The leaching out of flavolans correlates with the increasing relative consumption rates shown in the figure. This could, of course, be coincidental, but when flavolans are artificially added to otherwise palatable decayed *Ceriops* leaves, the amount eaten was markedly reduced. This confirms that flavolans deter feeding (Neilson *et al.* 1986).

Change and decay also alter the nutritional value of mangrove leaves. A useful rule of thumb measure is the ratio of carbon to nitrogen (C:N ratio). The higher this ratio, the less valuable the food. A C:N ration below 17:1 is regarded as being necessary for a nutritious food. Figure 4.5 shows that the C:N ratio tends to decrease with time of decay: the older the leaf, the better. Even at their best, however, rotting leaves appear to be an inadequate diet. The nitrogen-rich food items supplied by predation may form a small proportion of the total diet by volume, but may nevertheless be critical.

In areas where more than one species of mangrove tree occurs, crabs will have the choice of fallen leaves of different species, as well as leaves at different stages

Table 4.2 Consumption rate of *Neosarmatium smithi* when offered a choice of decayed, fresh or senescent leaves of *Ceriops tagal*, from Giddins *et al.* (1986)

Leaf type	Consumption rate g leaf wet weight per g crab per day ($\pm$ s.d.)
Fresh	0.004 $\pm$ 0.006
Senescent	0.011 $\pm$ 0.014
Decayed	0.062 $\pm$ 0.055

Fig. 4.5 Upper graph: tannin concentration (per cent of leaf dry weight: dashed line) and C:N ratio (continuous line). Lower graph: relative consumption rate by *Neosarmatium smithi* of leaves of *Ceriops* at different stages of decomposition. (Data, with permission, from Giddins, R.L. Lucas, S.J., Neilson, M.J. and Richards, G.N. (1986). Feeding ecology of the mangrove crab *Neosarmatium smithi* (Crustacea: Decapoda: Sesarmidae). *Marine Ecology Progress Series*, **33**, 147–155, © Inter-Research.)

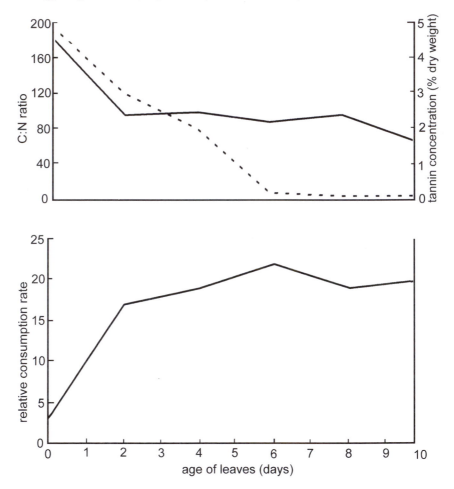

of decomposition. In some cases, crabs have been shown to prefer leaves of one species over those of another. Senescent leaves from three mangrove species were offered to two sesarmines common in the Mai Po marshes of Hong Kong, *Chiromanthes bidens* and *C. maipoensis*. The order of preference was *Avicennia marina*, followed by *Kandelia candel* then *Aegiceras corniculatum*. This correlates with the order of increasing C:N ratios: nutritionally, *Avicennia* is better value, followed by *Kandelia*, then *Aegiceras*. *Avicennia* also has the lowest content of aversive tannins (Table 4.3). An Australian species of sesarmine, *Parasesarma erythrodactyla*, has similarly been shown to prefer *Avicennia* to other species of mangrove (Camilleri 1989). Growth and survival of sesarmines appears also to

Table 4.3 Relative preference of two sesarmine species, *Chiromanthes bidens* and *C. maipoensis*, for leaves of three mangrove species compared with C:N ratios and soluble tannin content of the leaves, from Lee (1993)

Mangrove species	Preference rank	C:N ratio	% soluble tannins
Avicennia marina	1.30	27.4	0.86 ± 0.03
Kandelia candel	1.89	49.1	2.35 ± 0.29
Aegiceras corniculatum	2.80	69.1	1.95 ± 0.19

Preferences are mean ranking for both species and both sexes combined, and percentage soluble tannins are mean values standard deviation.

be better on a diet of *Avicennia* than *Kandelia*, and to be better on decayed leaves (with lower C:N ratios) than on fresh (Kwok and Lee 1995).

In contrast, studies in East Africa showed that *Avicennia marina* was the *least* preferred of five species of mangrove leaf, the sequence being *A. marina* < *Ceriops tagal* < *Rhizophora mucronata* < *Sonneratia alba* < *Bruguiera gymnorrhiza*. Two species of crab were involved: *Neosarmatium meinerti* and the gecarcinid land crab *Cardisoma carnifex*. Both showed the same order of preference, although only with *C. carnifex* were the results statistically significant. The experiment was carried out in high intertidal forests: it may be that the apparently perverse order of preference was due to the need to obtain scarce water from more succulent leaves, the water content of *Avicennia* leaves being relatively low (Micheli *et al.* 1991).

Although not all investigators have shown selective feeding, these experiments suggest that crabs may choose one species of fallen leaf over another, and in general prefer decaying leaves over fresh ones. In either case, the preference could be on the basis of tannin concentration or nutritional value.

Crabs do not always eat leaves on the surface: they may instead take them down burrows. After a week or two, tannin levels, and C:N ratios, fall to more acceptable levels. This behaviour makes sense from the crab's point of view, provided it does not abandon its burrow and contents in the mean time. In some areas, the amount of leaf material just below the mud surface suggests that crabs are not very efficient in dealing with buried leaves.

Sometimes leaves are eaten almost immediately after their disappearance below the surface, and before any appreciable change has taken place in tannin levels; one estimate was that 78 per cent of leaf material was eaten within 6 h of burial (Robertson 1988). Crabs may therefore take leaves underground for other reasons than to store them until they decay and become more palatable.

What other advantages might there be in taking food into burrows? Leaf burying, or caching, behaviour will sequester food resources, making them unavailable to competing crabs and other herbivores. Alternatively, burying leaves will prevent them from being washed out by the tide and extend the time when they can be fed on. If so, caching would be commoner where tidal outwash was more likely. Finally, burying leaves will minimise the time a foraging crab spends on the surface, exposed to predators or to low humidity, high temperature, or other uncongenial circumstances.

Leaf burial may therefore be commoner where there is greater predator pressure, or more intense competition between crabs for a limited supply of leaves, or under different tidal inundation regimes (Camilleri 1989; Giddins *et al.* 1986). A crab's life is a complex balance of opportunities and risks, costs and benefits, and how it allocates its time can be crucial to survival (p. 104).

Seedlings

Mangrove propagules or seedlings are also a major food source for sesarmine crabs. In Australia and Malaysia, the majority of propagules are destroyed within days of their release from the tree (and the released propagules are themselves the few survivors of pre-release attack by insects and other causes of death; see p. 56). Crabs show distinct preferences for different species of propagule. When propagules of five species of mangrove in tidal forests in north Queensland were tethered and their survival rates monitored over 18 days, there were clear differences in the rate of removal by crabs. As in the experiments on leaf litter, discussed above, crabs preferred *Avicennia marina*. Virtually every seedling of this species was destroyed by crab attack within the experimental period. Preference was again closely related to nutritive quality, *Avicennia* propagules having the highest content of simple sugars and lowest levels of tannins, fibre and protein (Smith 1987c).

The extent of seedling predation depends on other factors besides nutritional quality. Predation is generally greater at higher levels on the shore, reflecting shorter submergence times and consequently greater time available for crabs to forage (Fig. 4.6).

The greater the crab predation on a species, the less that species is likely to be a dominant constituent of the forest. This relationship is particularly striking in the case of *Avicennia*. In Queensland mangrove forests, *Avicennia* is present at the upper and lower shore, but absent from mid shore. *Avicennia* propagules do strand here, and there is nothing physically to prevent the species from estab-lishing itself: if crabs are artificially excluded it does so. The 18-day predation rate of tethered *Avicennia* seedlings in the midshore zone was 100 per cent, suggesting that the principal (if not the only) reason for the absence of *Avicennia* here is the presence of crabs (Smith 1987c).

Predation is also more intense in thick mangrove forest than where there are gaps in the canopy. The bigger the gap, the lower the rate of seedling predation (Fig. 4.7). This probably also relates to crab foraging behaviour. Either the extra travelling time in the open makes foraging less cost-effective, or it increases a crab's vulnerability to predators.

Although many other factors are no doubt involved, crabs appear to play a major role in litter processing, and in determining the local distribution of mangrove tree species and community structure of mangrove habitats.

Fig. 4.6 Cumulative predation of *Aegiceras* seedlings. Circles: low intertidal, squares: high intertidal, open symbols: open spaces, solid symbols: forest understorey. (From Osborne, K. and Smith, T.J. (1990). Differential predation on mangrove propagules in open and closed canopy forest habitats. *Vegetatio*, **89**, 1–6. Fig. 3 (part). © 1990 Kluwer Academic Publishers, reproduced by kind permission from Kluwer Academic Publishers.)

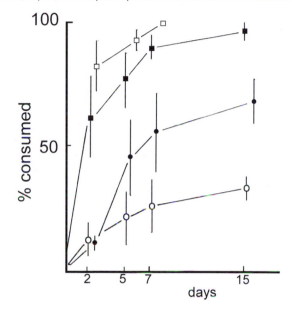

Fig. 4.7 Seedling predation by crabs (as per cent eaten) in relation to forest gap size. (From Osborne, K. and Smith, T.J. (1990). Differential predation on mangrove propagules in open and closed canopy forest habitats. *Vegetatio*, **89**, 1–6. Fig. 2. © 1990 Kluwer Academic Publishers, reproduced by kind permission from Kluwer Academic Publishers.)

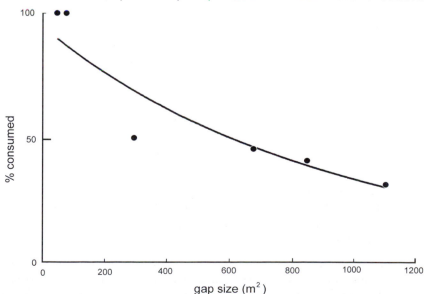

Are herbivorous crabs always important?

Overall, sesarmine crabs play a very important part in the ecology of man-groves, at least in Australia and Malaysia. Selective destruction of propagules determines the distribution pattern of tree species within the mangal, while processing of leaf litter expedites breakdown and makes this crucial energy source more readily available to other organisms. One estimate is that litter processing by crabs increase litter turnover to more than 75 times the rate that would obtain with microbial decay alone (Robertson and Daniel 1989). Are sesarmine crabs equally important in all mangrove ecosystems?

In the mangroves of south-west Florida, the crab fauna is dominated not by herbivorous sesarmines, but by deposit-feeding ocypodid crabs and predatory crabs belonging to the family Xanthidae. Experiments with tethered leaves, showed *no* removal by herbivorous crabs. There is some evidence that leaf-eating by crabs does take place in waterways after the leaves have been flushed out of the forest. Elsewhere in the Caribbean, leaf processing does take place: turnover decreases with height above low tide, the opposite of the situation in Australia. There appear to be important and interesting differences between the Caribbean and the Indo-Pacific in the ecological significance of mangrove crabs (McIvor and Smith 1996; Robertson and Daniel 1989).

Seedling predation by crabs also varies geographically. In Florida, overall rates of removal by crabs are, as would be expected, much lower: 6 per cent of *Avicennia germinans* seedlings were taken in Florida, compared with 95 to 100 per cent of *A. marina* in Australia and Malaysia. For the less desirable *Rhizophora*, the figures were similar in Australia and Malaysia: none were taken in Florida. Gastropod molluscs are more important herbivores than crabs in Florida mangroves, accounting for 73 per cent of seedlings, as opposed to 0 to 5 per cent in Australia and Malaysia.

The relationship between tree dominance and crab predation of seedlings has also been investigated in Caribbean mangrove communities. In Belize, preda-tion of *Avicennia*, as in Australia, was higher where the tree species was least dominant. The major seedling predators were the grapsid *Goniopsis* and the hairy land crab *Ucides* (Ocypodidae). In contrast, however, consumption of the seedlings of *Rhizophora mangle* and of a third mangrove species, *Laguncularia racemosa*, was highest where the adult trees were most, and not least, dominant (McKee 1995).

The importance of crabs in mangrove community structure and function there-fore differs from one part of the world to another.

Tree-climbing crabs

Crabs are not restricted to scavenging on the mud surface. Several species actually climb trees. In Malaysian mangroves, for example, the large sesarmine *Episesarma versicolor* climbs a metre or two up tree trunks, particularly at night.

These crabs appear to be scraping diatoms or other microphytes off the tree bark, but may also be avoiding aquatic predators such as fish or otters. They are certainly very vigilant where terrestrial predators are concerned, and drop off into the water if a human approaches closer than 2 or 3 m.

Episesarma is not truly arboreal. In East Africa, and in the Caribbean, are species which spend virtually all of their adult life in trees. The best known of these is the Caribbean species *Aratus pisonii* (Fig. 4.2). *Aratus* has short dactyls, used more or less like the pitons of a mountaineer. It is a very agile species and can achieve speeds of about 1 m s^{-1}. At night it moves up to the tops of trees, perhaps 10 m above ground, and feeds on buds and young leaves (Warner 1977).

An East African species, *Sesarma leptosoma*, virtually identical to *Aratus* in appearance, has a similar way of life. As an adult, this species never enters the water, nor does it even ventures onto the mud surface. Much of the day is spent foraging on the aerial roots of the host tree (usually *Rhizophora mucronata*, *Bruguiera gymnorrhiza* or *Ceriops tagal*) and, presumably, replenishing water. Twice a day crabs migrate in synchrony to the treetops. At about 06.00 h the 200 to 500 crabs occupying each tree move upwards, returning about 4 hours later. A second migration follows at around 17.00 h, with the return at dusk. While in the treetops the crabs feed by scraping tissue off the lower surface of mature leaves. They also spend much of their time on leaf buds at a particular stage of development. Besides food, the buds may accumulate valuable water. Individual crabs faithfully return to the same feeding sites by the same route, and return to the same root crevices on their return (Cannicci *et al.* 1996*a*, 1996*b*; Vannini and Ruwa 1994).

The remarkable synchrony of the vertical mass migrations may be to reduce predation. Fidelity to feeding and refuge sites may also give the crabs an advantage in avoiding predators through familiarity with the terrain. Alternatively, a fixed feeding area may improve the chances of finding patchily-distributed leaf buds. It is not clear, however, what particular combination of circumstances was conducive to the evolution of such complex behaviour in this particular species. Nor is it known how *S. leptosoma* copes with the problems of high tannin and C:N ratios which seem to have shaped the very different behaviour of other sesarmines.

Fiddler crabs

The most colourful mangrove invertebrates are crabs of the family Ocypodidae; among these, the most striking are undoubtedly the fiddler crabs (*Uca* spp.), which construct burrows at all levels of the shore. Fiddlers have bodies that are broad in relation to their length, and almost circular in midline section, so that they fit neatly, sideways, into a burrow. Males have one greatly enlarged claw, used in social displays and in jousting with rival males (Fig. 4.8). Particularly during displays, the carapace and claw are often very brightly coloured in

Fig. 4.8 Male *Uca* sp. (Photo: D. Barnes.)

crimson, orange or intense blue, depending on species. Females are much less flamboyant, and have two claws of normal size. Fiddler crabs have been encyclopaedically reviewed by Crane (1975).

Like many other members of their family, fiddler crabs are deposit feeders, ingesting organic matter from the exposed mud at low tide. Sediment is carried to the mouth parts by the minor chela in the male, and by both chelae in the female. The tips of the chelae are generally expanded, almost spoon-shaped, and often fringed with setae (Fig. 4.8).

Sediment is sorted within the buccal cavity. The outermost mouth parts, the third maxillipeds, play little active part other than helping to retain sediment and water during the sorting process. The inner surface of the second maxillipeds carries quite large numbers of long setae, some with spoon tips and others feathery. Facing these on the outer surface of the first maxillipeds is a brush of stiff setae. The sediment is rolled between two maxillipeds. The spooned setae of the second maxillipeds hold sand grains against the brush-like setae of the first maxillipeds, and diatoms and bacteria adhering to the grains are brushed off and moved towards the mouth itself. While this is going on, water is pumped out of the gill chamber into the buccal chamber. This helps the sorting process which takes place essentially in suspension. Sand grains, being denser, fall back towards the third maxillipeds. Together with unwanted organic matter, and some mucus from the crab's saliva, they are formed into pseudofaecal pellets and discarded. (These are described as 'pseudofaecal pellets', in contrast to faeces where waste material has passed through the gut before being discarded.) Much of the water is recirculated into the gill chambers, mainly through an aperture near the base of the outer, third max-illiped. Clogging of the gill surface with sediment particles is prevented by an array of filtering setae protecting this aperture.

Sediments vary greatly in particle size and organic content, and different species of *Uca* have specialised in different sediment compositions. Species which prefer coarser sandy sediments have chelae with a wider gape, to accommodate larger particles. Those which specialise in finer sediments tend to have a narrower gape, and to have the tips of the fingers of the claw fringed with setae to retain mud as it is carried towards the mouth (Fig. 4.9).

The mouth parts, too, are adapted to particular sediment compositions. Species feeding predominantly on coarse sandy sediments have more of the long, spoon-tipped setae on the inside of the second maxillipeds, and the tips of the setae are more spoon-shaped, while the setal 'brush' on the outside of the first maxillipeds is denser. Where the preferred sediment contains more fine organic particles, extra rows of setae are present at the base of the third maxillipeds to protect the aperture into the gill chamber and prevent the gills from becoming clogged (Macnae 1968; Miller 1961; Ono 1965).

Different food preferences are reflected also in the structure of the masticating structures in the foregut. Crabs have a single median tooth here, the surface of which carries a pattern of hard ridges (p. 84). Two lateral teeth, with rows of short stiff setae, are brought to bear on this surface. Species such as *Uca lactea* have a median tooth with parallel ridges, forming an extremely abrasive surface suitable for dealing with the coarse sandy sediments which this species prefers. In species which process finer, softer particles, the hard ridges on the median tooth are reduced and partly replaced by stout hooked setae, while the setae of the lateral teeth are longer and more flexible (Icely and Jones 1978).

The sophisticated sorting mechanism does more than just separate inorganic from organic particles. In two species of Australian fiddler crabs, *Uca vocans* and *U. polita*, bacteria and protozoa were found to be the most abundant cells in the gut, which contained relatively few algal cells. In the rejected material of the pseudofaecal pellets, the position was reversed, showing that the buccal sorting mechanism can select one type of cell and reject the other.

Selection continues within the gut. When the same two species of *Uca* were fed with radioactively labelled bacteria, radioactivity of the faeces was a tiny frac-tion of that of the gut contents, showing that bacteria were efficiently digested and not excreted. The assimilation efficiency was estimated at greater than 98 per cent in both species. In contrast, the assimilation efficiency for algal cells was 41 per cent and 31 per cent in *U. vocans* and *U. polita*, respectively. In general, deposit feeders such as *Uca* seem to depend much more on bacteria, diatoms, and the smaller members of the meiofauna such as ciliate protozoa and nema-todes than they do on larger algal cells and detrital material (Dye and Lasiak 1986, 1987; Jones 1984). Stable isotope analysis confirms that they do not depend on mangrove detritus, or do so only indirectly.

Fig. 4.9 Adaptations of fiddler crabs (*Uca* spp.) to different soil particle size. The three columns represent *Uca lactea*, *U. urvillei* and *U. dussumieri*, respectively. *Uca lactea* lives on coarser sediments than the other two species. First row (A): feeding chela (minor chela of male, either chela of female), second row (B): outer view of first maxilliped, third row (C): inner view of second maxilliped, fourth row (D): spoonshaped setae from second maxilliped, bottom row (E): outer view of third maxilliped. For description of how the mouth parts operate, see text. (Reproduced with permission from Macnae, W. (1968). A general account of the fauna and flora of mangrove swamps and forests in the Indo-West Pacific Region. *Advances in Marine Biology*, **6**: 73–270, © Academic Press; and Warner, 1977).

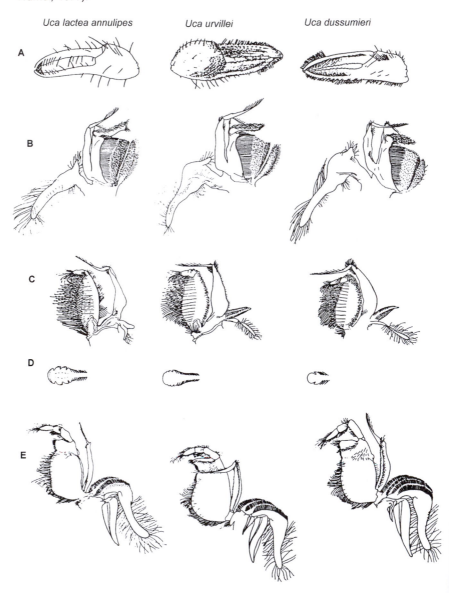

Because only a small proportion of the total sediment mass is selected as food, fiddler crabs must process large amounts of sediment. During periods of active feeding, the chelae are almost continuously active. The enlarged display chela of the males is useless for feeding, so males must work harder with the single minor chela, and have lower overall intake rates.

Poorer soils require a higher rate of intake: on a muddy Malaysian mangrove shore, rich in organic matter, male *Uca dussumieri* were recorded as feeding at the rate of 25 to 60 scoops per minute while *Uca lactea annulipes*, on a more sandy shore with lower organic content, scooped at a rate of 60 to 140 per minute (Macintosh 1988). Smaller and more local differences in soil organic content, detected by probing with the minor chela, may influence the behaviour of individual crabs. A calculating crab should abandon a particular patch of mud and move to another whenever its current rate of energy intake falls below a certain threshold of cost effectiveness. Measurement of feeding costs, in terms of oxygen consumption rates, and benefits, in energy value of the food acquired, makes it possible to predict this leaving threshold. The predicted threshold value will be different for males and females, as they have different intake rates and foraging costs, and at different concentrations of food in the substrate. At least one species of (salt marsh) fiddler crab investigated, *Uca pugnax*, behaves exactly as predicted. Fiddler crabs are therefore capable of optimising their foraging strategies to match a spatially heterogeneous environment (Weissburg 1993).

Most feeding takes place within a few hours of the tide receding, while the mud surface is still wet and easily handled. Later, when the surface has dried, is the time for more social activity. The allocation of appropriate time to different activities, as well as selection of where to forage, therefore promotes efficiency.

Fiddler crabs are often very abundant: densities of up to 60 per m^2 are not unusual in South-east Asian mangroves (Macintosh 1982). A single crab may produce around 300 pseudofaecal pellets a day, of around 36 mg dry weight each. Before the returning tide disperses them, pseudofaecal pellets sometimes almost entirely cover the soil surface. In addition, a fiddler crab may produce some 25 faecal pellets of 8 mg each. Compared with the unaltered substrate, the nitrogen content of these is enriched by a factor of approximately 5.6. Carbon content is also increased. Allowing for some of the material being processed more than once in the course of a day, the turnover rate of soil may be of the order of 500 g m^2 d^{-1}. Fiddler crabs therefore have a major role in retexturing the soil and altering its chemical composition, and in doing so significantly modify the soil environment (Macintosh 1982; Ono 1965). Burrowing also has a major physical impact on the environment, as discussed later (p. 105).

Fiddler crabs are also an important component of the mangrove ecosystem in another way: as food for numerous predators. These include mammals (racoons, monkeys), birds (kingfishers, herons), reptiles (snakes), fish (mudskippers) and other crabs. The annual production of fiddlers on a Malaysian mangrove shore has been estimated at 3.1 to 16.0 g m^2, equivalent to 4.6 to 28.9 kcal m^2, most of which is accounted for by predation (Macintosh 1982).

Reproductive adaptations

As with almost all species of crab, fiddler crabs have planktonic larvae. These, too, are a major food resource for predators, in this case mainly fish and other inhabitants of mangrove creeks and offshore waters. Crab larvae are minute, only 1 mm or so in length, but are copiously produced. One female *Uca rosea* (a common small South-east Asian species) has a brood size of 10 000: with around 4.5 batches a year, this gives an estimated annual production of 45 000 larvae. Taking into account the recorded population densities of fiddler crabs, and allowing for the proportion of mature females in the population, annual production may be of the order of half a million larvae per square meter of mangrove. A mangrove creek in Costa Rica, 325 ha in area, was conservatively estimated to export 7.5 10^{10} *Uca* larvae annually (Dittel *et al.* 1991; Macintosh 1982).

Females may synchronise the release of their larvae as a way of swamping predators and minimising losses, but only an infinitesimal fraction of the larvae released survive to settle on the shore and metamorphose into juvenile crabs.

Fiddler patterns in time and space

Mangrove fiddler crabs are active only during daytime, and when the soil surface is uncovered by the tide (Fig. 4.10). At other times, they retreat into burrows and remain inactive. Other considerations aside, therefore, it would be

Fig. 4.10 Pattern of activity of *Uca lactea* on a mangrove shore in Thailand, in relation to time of day and state of the tide. (Reproduced with permission from Macintosh, D.J. (1988). The ecology and physiology of decapods of mangrove swamps. *Symposia of the Zoological Society of London*, **59**, 315–341.)

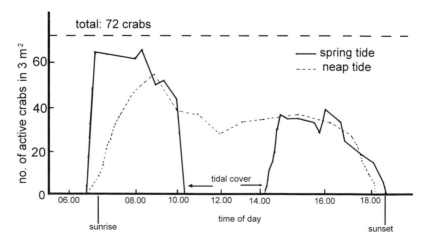

advantageous to live high on the shore since this allows more time for foraging and other activities. On the other hand, the particle-sorting method of feeding requires water, and although water is recirculated between buccal cavity and gill chamber, losses occur and must be replaced. Standing water is more readily available lower down the shore, so a crab foraging in the lower shore will not need to return to its burrow so frequently to replenish its water supply, and will be able to forage further afield.

The different species of *Uca* are adapted to feeding on different soil textures, and soil texture varies at different levels of the shore. Feeding methods therefore interact with other factors in determining the pattern of vertical zonation shown by fiddler crabs on a mangrove shore, an example of which is shown in Fig. 4.11.

Fig. 4.11 Vertical zonation of crabs on a Malaysian mangrove shore. Numbers shown for each vertical range are days per year without being covered by the tide. *Macrophthalmus, Uca* and *Ilyoplax* belong to the family Ocypodidae, the remainder are Grapsidae. EHWS, MHWS, MLWS, ELWS: extreme high water, mean high water, mean low water and extreme low water spring tides; EHWN, MHWN, MLWN, ELWN: the same, neap tides; MTL: mean tide level. (Reproduced with permission from Macintosh, D.J. (1988). The ecology and physiology of decapods of mangrove swamps. *Symposia of the Zoological Society of London,* **59,** 315–341.)

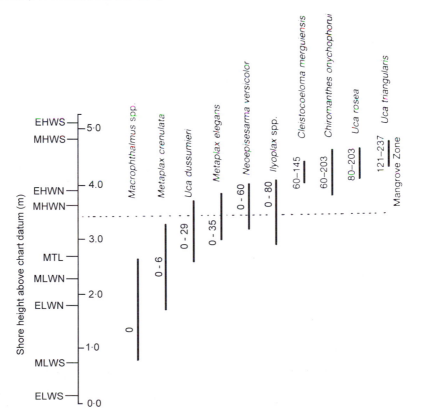

Other physiological problems which affect distribution are shared with sesarmines (and, indeed, with the majority of mangrove-living invertebrates), and are discussed next.

The physiology of living in mud

The problems facing mangrove crabs are much the same as those solved by mangrove trees themselves: coping with low oxygen concentrations in waterlogged mud, conservation of water and withstanding high temperatures and varying salinity.

Coping with low oxygen levels

Most mangrove crabs are active at low tide. While the tide is in and the mud surface submerged, they retreat into burrows and remain inactive. Fiddler crabs do not aerate the burrow by pumping water through it, so the oxygen concentration within the burrow is presumably as low as that in the surrounding mud. In laboratory conditions, *Uca* can survive anoxic conditions for up to 40 h. During this time the crab accumulates lactate within its body fluids. When it returns to normal conditions, oxygen consumption is raised for some time to eliminate the accumulated lactate. Fiddlers seem to survive anoxia by building up an oxygen debt, in a similar way to mammals (Warner 1977).

Sesarmine crabs presumably have similar tolerance of low oxygen levels, although this has not been investigated. Sesarmines caught overnight in flooded pitfall traps, in a hot Malaysian mangrove forest, are very often found dead. This may indicate limited tolerance of anoxia; or, perhaps, of high temperatures.

Most of the activity of fiddler crabs, and sesarmines, takes place in air. The oxygen concentration of air, at 20 per cent, is of course considerably higher than that in water, even at saturation levels. Offsetting this advantage is the fact that gills in general are not well adapted to air breathing. Finely-divided respiratory tissues tend to collapse under surface tension, reducing their effective surface area for gas exchange. Accordingly, the gill area of *Sesarma* and *Uca* is very much reduced compared with that of fully aquatic crab species, and the walls of the branchial chamber are vascularised for gas exchange, in an approximation to a vertebrate lung (Gray 1957; Takeda *et al.* 1996)

Sesarmine crabs have a supplementary mechanism. The ventral carapace plates on either side of the mouth parts and below the eyes carry a very distinctive array of short bristles. Water is pumped out through exhalant openings at the side of the buccal cavity, is spread as a thin film by the network of bristles, and is then taken back into the gill chambers again at an inhalant opening near the base of the legs (Fig. 4.12). Gas exchange with the atmosphere takes place during this recycling process. Some water passes over the crab's dorsal surface, where it is again spread out into a thin film by lines of short bristles. Here, too, some gas exchange occurs (Felgenhauer and Abele 1983).

Fig. 4.12 The reticulated network of short setae, and the route of water recirculation, on the under surface of a sesarmine crab, *Sesarma reticulatum*. 1.Frontal view of the crab. White and black arrows indicate the path of respiratory water across the carapace (scale bar = 1 mm). 2. Close-up views of the seta-covered plate to show arrangement of setae (scale bar = 100 µm). 3. Further magnification to show presence along shaft of seta of fine setiules for water retention (scale bar = 50 µm). (Reproduced with permission from Felgenhauer, B.E. and Abele, L.G. (1983). Branchial water movement in the grapsid crab *Sesarma reticulatum* Say. *Journal of Crustacean Biology*, **3**, 187–95. © 1983 by The Crustacean Society.)

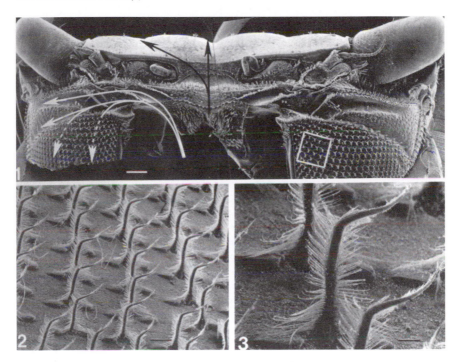

High temperature

The temperature at the mud surface in a tropical mangrove swamp can rise to 45°C, close to the lethal limit for most animals. Evaporation lowers the body temperature by a critical few degrees: fiddler crabs, for example, can keep their body a few degrees below the prevailing air temperature. The external recirculation of water by sesarmines probably cools the crab by evaporation, as well as facilitating gas exchange. For the most part, however, mangrove crabs probably maintain a reasonable body temperature by movement in and out of burrows. Figure 4.13 shows the temperatures measured at various locations in which a fiddler crab (*Uca lactea*) spends its time. As the temperature rises, a crab makes more frequent trips back to its burrow and spends proportionately less time on feeding and other activities (Edney 1961; Eshky *et al.* 1995).

Fig. 4.13 Temperature at various points occupied by a fiddler crab (*Uca*). (Reproduced with permission from Edney, E.B. (1961). The water and heat relationships of fiddler crabs (*Uca* spp.). *Transactions of the Royal Society of South Africa*, **36**, 71–91.)

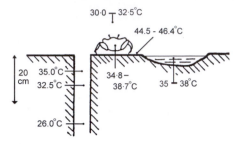

Water economy

It may seem strange to suggest that animals living in a mangrove swamp might have a problem with water shortage. In some mangrove areas, the tide goes out a considerable distance horizontally, and pools of standing water are not necessarily abundant. Frequent returns to a burrow to replenish water take time that could be devoted to feeding or social behaviour.

Some species are relatively impermeable to water. Water passes through the carapace of *Uca* at rates of about 3 mg cm^2 h^{-1}, and *Sesarma* at 6 mg cm^2 h^1, compared with a range of 14 to 21 mg cm^2 h^{-1} for a variety of non-mangrove species (Vernberg and Vernberg 1972, quoted in Hutchings and Saenger 1987). This reduction in water loss through the carapace will be at the expense of evaporative cooling.

Some species, including *Sesarma*, have an additional means of gaining water from the soil by what are, in effect, roots. Tufts of hydrophilic setae at the bases of the legs are brought into contact with the moist surface of the mud and can actually draw water into the crab's body (Burggren and McMahon 1988).

Osmotic problems

The final problem to consider is that of varying salt concentrations, specifically, the osmotic imbalances that might result. Crabs, of course, are marine in origin: their starting point is therefore the salinity of full sea water, in contrast to mangrove trees which are of terrestrial origin and come from ancestors accustomed to fresh water.

Only a few osmoregulatory strategies are possible. An animal could be an osmoconformer and allow its body fluids simply to equilibrate in osmotic concentration with the external medium. This saves energy that would otherwise be devoted to osmoregulation by actively transferring water in or out of the body. Perfect osmoconformity over the full range of salinities is impossible: at very low external concentrations, tissue fluids and cell contents would have to approach the composition of pure water, while extremely high ionic concentrations interfere with protein conformation and physiological functions.

Perfect osmoregulation would be expensive in a small animal (with a high surface area/volume ratio), without a completely impermeable integument. This extreme strategy is therefore not available to crabs.

Between these extremes, various compromises are possible. An animal could maintain an osmotic concentration greater than that of the environment (hyperosmoregulation) at low osmotic concentrations, but conform at higher ones, or hyperosmoregulate at lower concentrations and hyporegulate at higher ones.

Among mangrove crabs, the commonest strategies involve hyperosmoregulation at low external concentrations, and a varying degree of hyporegulation at higher concentrations (Fig. 4.14).

Fig. 4.14 Osmotic concentration of the blood of some ocypodid crabs in varying external osmotic concentrations. Osmotic concentration is given as freezing point depression (Δ°C); the diagonal line represents complete osmotic conformity with the external medium. SW = sea water, approximately 35‰. (Reproduced with permission from Barnes, R.S.K. (1976). The osmotic behaviour of a number of Grapsoid crabs with respect to their differential penetration of an estuarine system. *Journal of Experimental Biology*, **47**, 535–51, copyright Cambridge University Press.)

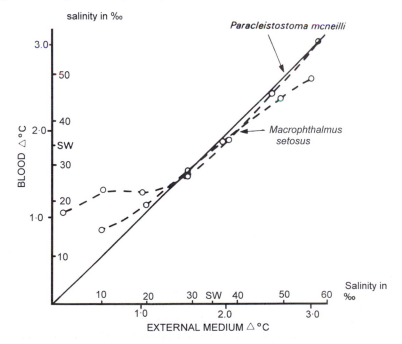

How stressful is the life of a crab?

If a crab is turned on its back and released, it soon struggles back to the correct position and scuttles off. The healthier the crab, the quicker it is to do this. If the crab has been suffering some form of physiological stress, its physiological responses are slower and co-ordination poorer, and it is slower to right itself. Measuring the righting response time (RRT) can therefore give a quantitative assessment of the degree of physiological stress to which a crab has been subjected.

This method has been used to judge how stressful conditions in mangrove habitats are to crabs, by measuring the RRT after crabs had been restrained for periods of 3 h on the mangrove mud, among mangrove roots, or in the mangrove canopy. Controls were freshly caught crabs. By this criterion, the most stressful situation was in the mangrove canopy, followed by the mangrove roots, then the open mud. Least stressed were the freshly-caught controls (Wilson 1989). This is not surprising, although it would be interesting to compare stress levels in the canopy for a habitually tree-climbing species (p. 92). It does raise the question of why crabs ever leave the least stressful environment of open mud. Again unsurprisingly, crabs are more exposed to predation on the open mud than they are amongst mangrove roots. Physiological stress and predation risk are being traded off against each other.

A crab's life is one of compromise and trade-offs: of one physiological stress against another, of predation risk against physiological stresses, of the need to feed efficiently, and social and reproductive requirements against all of these. Particular patterns of distribution and activity in time and in space are the resultant of these conflicting requirements.

Other mangrove crustacea

Grapsid and ocypodid crabs, particularly *Sesarma* and *Uca* species, dominate the mud surface at low tide. As we have seen, these are predominantly herbivores and deposit-feeders, respectively, although some members of the families are relatively omnivorous. Other crabs found in mangroves may be important predators. The most conspicuous is the mud crab *Scylla serrata*, of the family of swimming crabs (Portunidae). *Scylla* reaches a carapace width of up to 20 cm, making it the largest invertebrate predator found in mangroves. Equally formidable predators are the mantis shrimps (Stomatopoda) which live in burrows in the mud, and lacerate prey by rapidly shooting out their spiked 'raptorial appendages': at least one species, *Squilla choprai*, is found in mangroves, in Malaysia (Macnae 1968).

More general mangrove scavengers include hermit crabs, particularly *Clibanarius* species, which forage on the mud surface at high tide. Some hermits may also climb into the mangrove trees. Shrimps and prawns (the terms are used interchangeably) may also be abundant in mangroves and mangrove creeks. Penaeid shrimps, which in at least some parts of the world depend heavily on mangroves for feeding and breeding, are an important commercial crop. The relationship of mangroves to shrimp fisheries and aquaculture is discussed later (p. 144).

The shrimp *Merguia* apparently lives only in mangroves, and has the distinction of being the only semiterrestrial shrimp known: it actually climbs trees. Only two species are known. One occurs in the Indo-West Pacific region, from Kenya to Indonesia; the other has been found in Panama, Brazil and Nigeria (Bruce 1993). Indo-West Pacific and Atlantic regions differ in the composition of their mangrove floras (see p. 152). The separation of the two species of mangrove-associated shrimps presumably paralleled the divergence of the mangroves themselves.

Numerous crustacea live under the mud surface. Pistol shrimps (*Alpheus*) are more often heard than seen: they live in burrows in the mud, and have one greatly enlarged claw, used to make the loud snapping sound which gives them their common name. Also seen only rarely at the surface is the mud lobster (*Thalassina*), a subterranean deposit feeder which ejects waste mud into sizeable mounds (Macnae 1968).

Crustacea as ecosystem engineers

All species have an impact on their environment: at the very least, they exchange materials in the form of food, waste materials and respiratory gases. Some species have an effect beyond these straightforward transactions, and alter the nature of the environment in ways that affect species other than their direct competitors, predators or prey. Such species are often termed 'ecosystem engineers'. Ecosystem engineers are species that 'directly or indirectly modulate the availability of resources to other species, by causing physical state changes in biotic or abiotic materials. In so doing they modify, maintain and create habitats' (Jones *et al.* 1994).

The trees themselves are the greatest of the ecosystem engineers in mangroves. Self-evidently, they create the mangrove habitat, and by their physical nature and organic production greatly modify the muddy soil in which they grow. Mangrove crustaceans, and particularly crabs, are also significant ecosystem engineers. The topography of mangrove swamps in South-east Asia is often visibly modified by mud lobsters (*Thalassina*). In the processing of burrowing, and extracting organic matter from the mud, *Thalassina* throw up waste material from beneath the surface. This accumulates as mounds which are often well over 1 m in height, and may even reach 2 m. *Thalassina* mounds create large patches of relatively dry mud which provide suitable habitats for a variety of other species, including the mangrove fern *Acrostichum*, fiddler crabs and the large sesarmine *Episesarma*, as well as other burrowing crustacea and molluscs. Between the mounds the mud is more frequently inundated by the tides, and more waterlogged, than it would otherwise be. Burrowing crabs also contour their environment, although not so dramatically (Warren and Underwood 1986).

Crab activities also have less dramatic effects. Much of the microbial activity of mangrove mud occurs within a short distance of the surface, to a depth limited by diffusional gas exchange with the atmosphere. This activity will be enhanced by fiddler crabs working over the surface as they extract food, continually exposing fresh mud to the surface and sorting soil particles by size and composition. Burrowing has the further effect of increasing the surface area of mud exposed more or less directly to the atmosphere. A square metre of mangrove mud may contain 40 to 50 crabs. If each crab constructs a burrow 10 cm deep and 1 cm in diameter, the result is to increase the surface area of the mud by around 12 per cent (quite apart from the effects of redistributing the material excavated from the burrows during construction). This simple calculation may underestimate the importance of crab burrows: an analysis of the effects of fiddler crabs in a salt marsh indicated that burrowing increased the surface area by 59 per cent (Katz 1980).

Although some crab burrows are relatively simple in structure, with exclusive occupancy by a single crab, others are not. Mangrove mud is often honeycombed with a network of interconnecting passages, more or less promiscuously occupied by a shifting population of crabs. The labyrinthine state is rarely obvious from the surface of the mud, but is revealed at low tide by the amount of water that flows out of crab holes in the banks of mangrove creeks, and by the observation that the rising tide often wells out of crab burrows long before it starts advancing across the surface of the mud. Dye injection confirms the extent of water movement under the mud surface, through crab burrows (Ridd 1996).

Are effects such as these of any importance? Fiddler crabs are known to influence the productivity of salt marsh vegetation: reduction in crab density in small experimental plots resulted in a 47 per cent decrease in leaf production of the salt marsh grass *Spartina* in a single season (Bertness 1985). Removal experiments are less easy to carry out in mangroves, but show similar effects. The removal of crabs from 15 m × 15 m enclosures within Australian mangroves resulted in a significant increase in soil sulphide and ammonium levels, and a reduction in cumulative growth and reproductive output of the trees (Smith *et al.* 1991).

Burrowing crustaceans are, therefore, significant ecosystem engineers. Not only do they alter the physical structure of their environment, but they also significantly affect the growth and productivity of the mangrove trees.

Mangrove molluscs

Snails

The most conspicuous mollusc inhabitants of mangroves are gastropod snails. As with the crustacean fauna, these include deposit feeders, herbivores which browse on fallen leaves or on algae growing on tree bark and predators.

Although there are a few species unique to mangroves, most of the species that live on mangrove mud are also found on open mudflats without mangroves. Species that live on trees are generally also found on other hard surfaces, such as rocks. Very few species are exclusive to mangroves.

The principal predatory gastropods are species of *Thais*, found in mangroves world-wide. These cruise over mud and mangrove roots, feeding on barnacles or small gastropods. *Thais* drills through the shells of its prey with the radula, a ribbon-like tongue covered in numerous sharp teeth, which operates rather like a miniature chainsaw.

Most of the snails found on the mud are deposit feeders, using their radula to scrape organic particles from the surface. Some genera, such as *Cerithidea*, have representatives in mangroves throughout the world, others are geographically more restricted (Plaziat 1984). Among the more spectacular are species of *Telescopium* and *Terebralia*, widespread in the Indo-Pacific region. The cone-shaped shells may reach 190 mm in length. Snails of this size are probably virtually exempt from predation. Crushed *Telescopium* shells are sometimes found scattered around burrows of the large portunid crab *Scylla*, but no other invertebrate predator could cope. Unfortunately the large size of this species has attracted human attention, and in many places it is over-exploited, as food or for other purposes. In the Vellar estuary, India, several tonnes of *Telescopium* are collected annually as a source of calcium carbonate for the lime industry (Houbrick 1991).

Few gastropods eat mangrove leaf material directly. One which does is *Terebralia palustris*. This species changes its diet with age. Young *T. palustris* are detritus feeders. On reaching a shell length of around 30 mm, they begin to cluster on fallen mangrove leaves. In small specimens, the radula carries denticulate teeth, virtually identical in size and shape to the specialised spoon-shaped setae of fiddler crab mouthparts (Fig. 4.9). This remarkable example of convergent evolution reflects a close similarity in function: in both cases, the structure is used to separate fine organic particles from inorganic matter. As *Terebralia* grows, the radula teeth metamorphose to the adult pattern becoming relatively larger, and with simple blade-like cutting edges. Examination of gut contents shows that pieces of leaf material are being ingested.

Further evidence that large *Terebralia* feed almost exclusively on mangrove leaves comes from studies of the isotopic composition of the snails. Adult *Terebralia palustris* tissues show $\delta^{13}C$ values lower than $-21‰$, close to those of mangrove leaves and in contrast to the values found in species known to have other diets: the deposit-feeding fiddler crab *Uca lactea*, for instance, having a value of $\delta^{13}C = -18.96$. This is good evidence that adult *L.palustris* feeds primarily on leaf material, and can be a major means of its removal from the forest floor. Juveniles show values in the range -17.5 to $-18.5‰$, and show a transition to adult $\delta^{13}C$ values which correlates with metamorphosis of the radula (Fig. 4.15) (Marguillier *et al.* 1997; Slim *et al.* 1997).

The most abundant snails on mangrove trees are species of *Littoraria*, close relatives of the periwinkles of temperate rocky shores. In Central American

Fig. 4.15 Change in δ^{13}C of tissue of *Terebralia palustris* with increasing shell length, indicating a dietary change from detritus to mangrove leaves. (Reprinted from Slim, F.J., Hemminga, M.A., Ochieng, C., Jannink, N.T., Cocheret de la Morinière, E. and van der Velde, G. (1997). Leaf litter removal by the snail *Terebralia palustris* (Linnaeus) and sesarmid crabs in an East African mangrove forest (Gazi Bay, Kenya). *Journal of Experimental Marine Biology and Ecology*, **215**, 35–48, © 1997, with permission from Elsevier Science.)

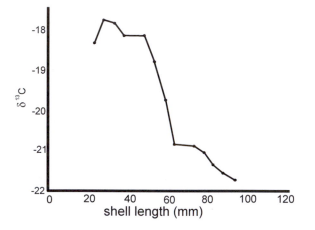

and Caribbean mangroves, on both sides of the Isthmus of Panama and in West Africa, the common species is *L. angulifera*. In the Indo-Pacific, this species is replaced by a number of others, which partition between them the slightly different habitats available on the surface of mangrove trees. In Papua New Guinea, *Littoraria scabra* and *L. intermedia* are found only on the bark of trunks, roots and branches. *L. scabra* is commoner on trees at the seaward edge of the mangrove forest, while *L. intermedia* prefers trees facing onto freshwater creeks. The shell colour of these two species closely matches that of the bark on which they live (as does that of other bark-living snail species). The mouth of the shell is rounded, to fit the contours of root surfaces, and the radula relatively long as an adaptation to the rough surface of the bark.

The leaves are occupied by a different species, *L. pallescens*. This species has a flat aperture to match its substrate, and a much thinner and weaker shell. Shell-crushing crabs forage on roots and even trunks of mangrove trees, but not on leaves. Leaf-living snails have less need of protection against mechanical damage by crabs, and can afford to have thinner shells. *Littoraria pallescens* is conspicuously polymorphic. Typically, up to 9 per cent are orange or pink, 10 to 50 per cent yellow and the remainder dark in shell colour.

Stable polymorphisms such as this often result from a balance between conflicting selection pressures. Some colour varieties may be more conspicuous than others to a predator such as a bird, or predators may prefer whichever variety is the most frequent in the population. The physical environment may also affect polymorphism: yellow *L. pallescens* gain less heat from sunlight than the dark forms, and are typically around 1.5°C cooler, with the orange form intermediate.

The polymorphism presumably results from a balance between such physical and biotic factors (Cook 1986; Cook and Freeman 1986; Cook *et al.* 1985).

Like crabs, mangrove snails have to survive varying salinity, high temperature and anoxia, and avoid desiccation. Of these physical problems, salinity may be the least serious. Most mangrove snails are probably quite tolerant of a range of salinities: *Terebralia palustris*, for example, can live comfortably in salinities of 15 to 35 ppt (Houbrick 1991). The connected issues of heat and water conservation are probably more limiting. There is little direct information, but the clustering of many species of snail in shady positions among mangrove roots suggests the avoidance of heat stress and accompanying water loss.

Like crabs, most mangrove gastropods feed in damp conditions at low tide, and are therefore adapted to aerial respiration. In many species the gills are therefore reduced (*Terebralia palustris* being a surprising exception) and gas exchange takes place over the folded and vascularised epithelium of the mantle cavity (Plaziat 1984).

High-shore species, in areas that dry out for long periods, tend to become inactive to conserve water. Closing tightly to minimise evaporation also limits gas exchange, necessitating periods of anaerobic respiration. This creates acidic conditions. Some species, particularly those found in habitats rich in detritus, appear to counter internal low pH conditions by the dissolution of calcium carbonate from their shells. Often this results in the thinning of the apex of the shell to the point where it is completely lost. This process, known as decollation, is characteristic of South-east Asian species such as *Cerithidea*, but has not been reported from comparable species in West Africa or Panama (Vermeij 1974).

The external environment, of course, also tends to be anoxic and of low pH, and exterior shell erosion can be a problem. The shells of epifaunal molluscs are protected by a thick outer non-calcareous periostracum: if this is scratched, erosion of the underlying shell can occur (Plaziat 1984).

Bivalves

The most visible bivalve molluscs of mangroves are the encrusting oysters and mussels found attached to roots (see p. 78). Within the mud, burrowing filter-feeding bivalves may be abundant. Infaunal species tend to be found also in adjacent open mudflats, so are not peculiar to mangroves. Stable isotope analysis indicates that they do not depend directly for food on particles of mangrove detritus, but either on algal cells or on organic particles processed by other organisms.

Bivalve molluscs are also important components of the community that lives within mangrove wood, particularly in dead wood. There is some evidence that wood-boring animals can colonise wood only after it has first been attacked by fungi. Typical wood-boring molluscs are the shipworms of the family Teredinidae. (Despite their common name, shipworms are bivalve molluscs, not worms.) These include the giant mangrove shipworm *Dicyathifer*, which may reach 2 m in length; most are of more normal proportions (Plaziat 1984).

Meiofauna

Most studies of mangrove fauna consider animals visible to the naked eye, whether on the mud surface or beneath it. Like other soft sediments, mangrove mud also contains an abundant and diverse meiofauna, comprising animals small enough to live between soil particles. Figure 4.16 shows some representative meiofaunal organisms. The few studies of mangrove meiofauna that have so far been carried out have shown a little of its interest and importance.

Meiofauna are not easy to study. The organisms, besides being tiny, are often transparent and are easily damaged by being physically extracted from the mud. Extraction involves inserting a tube 2 to 3 cm in diameter into the mud, dispersing the core of mud obtained in a suitable liquid, and adding a stain such as Rose Bengal to make visible some of the more translucent animals.

Fig. 4.16 Representative meiofaunal organisms. (a) nematode (total body length approx. 1 mm); (b) a harpacticoid copepod (*Prionos ornata*) (scale bar 100 µm); (c) turbellarian (scale bar 200 µm); (d) gastrotrich (scale bar 100 µm); (e) kinorhynch (scale bar 100 µm); (f) ciliate protozoan (scale bar 100 µm); (g) tardigrade (scale bar 50 µm); (h) harpacticoid copepod (*Lizashtonia hirsutosoma*: see also Fig. 5.1) (scale bar 100 µm). ((b) and (h) come from mangrove sediments in western Peninsular Malaysia, and are reproduced by courtesy of Rony Huys and Mike Gee. (c)–(g) reproduced with permission from Higgins, R.P. and Thiel, H. (eds) (1988). *Introduction to the study of meiofauna*, Smithsonian Institution.)

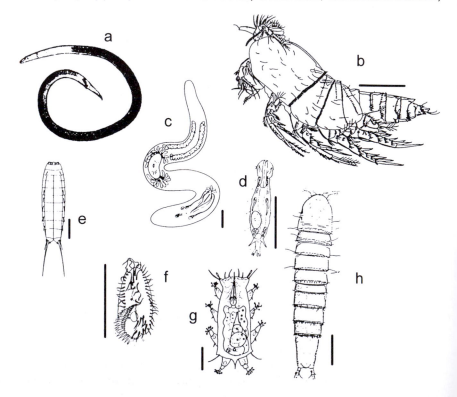

Then, because mangrove swamps are usually some way away from laboratories with the necessary microscopes, the samples are preserved with formalin for later analysis. Usually, this consists of separating out meiofaunal animals by sieving; meiofauna being defined, to some extent arbitrarily, by the mesh size of available sieves. In one study, meiofauna were regarded as those organisms that passed through a mesh of 1000 μm, but were retained by a mesh of 53 μm (Sasekumar 1994). Some soft-bodied organisms, such as minute turbellarian flatworms, disintegrate during this process, so are underestimated when samples are analysed; and differences in the methods used by different investigators make it difficult to compare results from different sites. Nevertheless, some interesting patterns are emerging.

As a general rule, small animals are more numerous than large. It is no surprise, then, that beneath a 10 cm^2 area of mud there may be many thousands of individuals. Most of these live in the upper centimetre or so of the mud, with numbers falling off rapidly below this depth (Somerfield et al. 1998). Meiofaunal populations are also more diverse than the macrofauna: not only are there many species, but the species show a higher level of taxonomic diversity. The macrofauna is dominated by two phyla, molluscs and arthropods, with only a few other phyla being represented, while the meiofauna includes representatives of many phyla. One set of meiofaunal samples from the mangroves of Hinchinbrook Island, Australia, for example, yielded 1600 turbellarian flatworms, 200 nematodes, 9 harpacticoid copepods and sundry ciliate protozoa, foraminifera, bivalve molluscs, oligochaetes, polychaetes, hydrozoa, archiannelids, kinorhynchs, amphipods, cumaceans and other crustacea, tardigrades and gastrotrichs (Alongi 1987a).

The abundance and composition of the meiofauna vary with the nature of the soil, with sediment depth, shore level with the dominant tree species and seasonally. In Kenya, the greatest abundance was in the *Bruguiera* zone, followed by *Rhizophora*, *Avicennia*, *Sonneratia* and *Ceriops* (Vanhove et al. 1992). In contrast, in Malaysia relative densities were reversed: the greatest abundance was in the *Avicennia* zone, followed by *Rhizophora*, then *Bruguiera* (Sasekumar 1994). It is not clear whether procedural differences explain such disparities, or whether genuine geographical differences are involved.

It is quite possible that meiofaunal populations also show variation at much smaller scales. If macrofaunal animals vary in relative abundance over horizontal distances of the order of a few metres, meiofaunal animals may vary in abundance over distances of the order of a few centimetres. Sampling methods used so far would have no chance of picking up variation on this scale.

Little is known of the food chains or community structure of the meiofauna. As with the macrofauna, the meiofauna operates at a number of trophic levels. There are meiofaunal herbivores, predators and detritivores just as there are at much greater body sizes. Heterotrophic bacteria are probably a major food resource. Rates of bacterial production, at least in the Hinchinbrook Island mangroves mentioned above, are typically much greater than the maximal ingestion rates of the meiofauna, so the primary consumers among the meio-fauna are not limited by food availability. High tannin levels leaching out from mangrove litter may depress meiofaunal abundance (Alongi 1987b). Variation in the tannin content in the leaves of different species may explain some of the initial differences in the meiofauna colonising newly fallen mangrove leaves. As decay proceeds, the meiofaunal community undergoes a series of successional changes, reaching a climax state after a few weeks which is little affected by the species of leaf. The dominant organisms at this stage are nematodes and copepods (Gee and Somerfield 1997).

Does the meiofaunal microcosm interact significantly with macrobenthic organisms whose world is on such a different scale? The meiofaunal food web seems relatively independent, with only minor energy fluxes to the larger epibenthos. Interactions between meiofauna and the benthic macrofauna have to some extent been clarified by experiments in which the latter—particularly crabs (*Uca*) and molluscs such as *Terebralia* and *Cerithidea*—are excluded by cages from areas of mangrove mud. The results depend on area, sediment type, time of exclusion and other factors, but in general the consequence is an increase in abundance of meiofauna.

This could be due to the elimination of predators allowing prey numbers to increase. However, predation of meiofauna by macrofauna seems to be relatively unimportant. Although fiddler crabs process a high proportion of the surface layers of sediment, they selectively remove bacteria and ciliates, and other organisms are mostly returned to the mud in pseudofaecal pellets (see p. 94). Harpacticoid copepods are eaten by juvenile fish and crabs, so might be expected to increase in numbers during an exclusion experiment. Unlike other groups, copepod abundance actually tends to decrease in exclusion experiments, and in equivalent laboratory microcosms (Olafsson and Ndaro 1997; Schrijvers and Vincx 1997).

The meiofaunal groups which show the greatest increase in abundance in long-term exclusion experiments (several months to a year) are nematodes and oligochaetes. The species in which the increase in numbers was most marked were those known to feed on microalgae or fine bacteria-rich organic matter. Neither nematodes nor oligochaetes are significantly eaten by crabs, so the increase in numbers is not likely to be due to relaxation of predation pressure. Despite the disparity in size, these minute organisms compete for food with the much larger epibenthic macrofauna (Dye and Lasiak 1986; Schrijvers and Vincx 1997; Schrijvers et al. 1997).

Fish

The creeks, pools and inlets of mangrove forests typically have a rich fauna of fish. With such mobile animals, abundance is impossible to measure directly, although mangrove fish are often the basis of thriving local fisheries (see p. 148). Species richness is generally high. In the mangrove creeks of Selangor, western

Malaysia, one survey recorded 119 species. In terms of individual numbers, 70 per cent of the sample comprised schooling fish such as anchovies; most of the biomass was made up of larger catfish, mullets and archerfish. Almost all were juveniles, suggesting that the mangrove habitat serves as a 'nursery' area for the surrounding waters. The mullets (*Liza* spp.) consume significant quantities of mangrove detritus; other species generally eat mangrove invertebrates including crabs and their planktonic larvae, gastropod and bivalve molluscs and sipunculids (Sasekumar *et al.* 1992).

How typical are the mangroves of Selangor? Figures from other mangrove areas include 83 species in Kenya, 55 and 133 from two areas of Queensland, Australia, 59 species in Puerto Rico and 128 from the Philippines (Chong *et al.* 1990). Estimates of diversity depend heavily on catching methods and intensity, so these figures are not directly comparable. However, it can probably be concluded that mangrove creeks generally have a rich fish fauna.

Fish caught in creeks presumably forage among the mangrove trees at high tide. At low tide, this habitat is occupied by the mudskippers, relatives of the gobies (Fig. 4.17). Anyone who has visited mangroves or tropical mudflats in the Indo-Pacific region will have been captivated by these fish. As their common name suggests, they skip across the surface of the mud by using their tail, and their pectoral fins which serve almost as legs. Some even climb up aerial roots, where they cling on partly by grasping the root with their pectoral fins, as in many species the pelvic fins are fused together to form a sucker.

Fig. 4.17 Mudskipper on a pneumatophore of *Avicennia*. (Photograph HAR.)

As the tide rises, mudskippers commute upshore (or up trees), and as it falls they move back down across the mud. Despite this mobility, they are generally territorial, sometimes maintaining high shore burrows as well as low shore feeding territories, in the core of which is a further burrow system. Territories are defended against conspecifics and other intruders such as crabs, and are the basis of courtship and other social interactions. Social displays involve a complex series of postures, erection of the conspicuous dorsal fin, undulations of the tail, head-butting and biting (Clayton 1993; Macnae 1968).

Burrows vary in structure, but usually include a more or less vertical shaft with horizontal side tunnels. Sometimes the burrow is shared with a commensal pistol shrimp (*Alpheus*) or crab. Regular commensal associations between gobioid fish and alpheid shrimps are well known in other habitats, such as sandy shores; some of the other associations may be more fortuitous, resulting from the accidental anastomosis of mudskipper and crab burrows.

A mudskipper burrow may have surface embellishments, such as carefully constructed mud towers or rimmed saucer-like depressions. These might have a display or other social function, although a tower at one end of a U-shaped burrow would also promote water movement through the burrow as a horizontal tidal current washed past.

Mudskippers of the genera *Periophthalmus* and *Periophthalmodon* are predominantly carnivorous and, like many carnivores, often unselective in their prey. The diet of *Periophthalmus* includes crabs, insects (particularly ants), spiders, nemertine worms, shrimps, copepods, amphipods and snails. *Periophthalmodon* appears to specialise in crabs, particularly sesarmines and fiddler crabs.

In contrast, *Boleophthalmus* is largely a deposit feeder, skimming sediment from the mud surface with a peculiar sideways movement of its head. It then moves the mouthful of food and water around the buccal cavity, possibly sorting it into its constituents in a similar way to *Uca* (p. 94), before squirting water out through its mouth and gill apertures. Other species, such as *Scartelaos*, have a more omnivorous diet.

All mudskippers are probably to some extent omnivorous when the situation demands, and there are certainly differences in diet between small and large individuals of the same species. The differences in the predominant diet, however, are sufficiently consistent within species to be reflected in morphological adaptations of the gut. As a general rule, plant material is less easily digested and herbivores therefore tend to have longer guts. This is shown by the relative gut lengths of mudskipper species, shown in Table 4.4.

Mudskippers have appropriate physiological adaptations to the demanding amphibious way of life. As long as they can return periodically to burrows, water loss is probably not a major problem to mudskippers. Some species, however, do have considerable ability to survive desiccation where necessary. Although it rarely spends long periods of time out of water in normal circumstances, the Chinese mudskipper *Periophthalmus cantonensis* can survive 2.5 days

Table 4.4 Relative gut length of mudskippers, data quoted in Clayton (1993)

Species	Feeding mode	Relative gut length			
Periophthalmus	carnivore	0.5–0.6	0.64–0.76	0.39	0.28–0.45
Periophthalmodon	carnivore	1.09	0.43	0.8	0.42
Scartelaos	omnivore	1.27			0.6
Boleophthalmus	herbivore	2.55	1.45		
Apocrytes	herbivore				2.0
Pseudoapocrytes	herbivore				2.7

Because guts are difficult to measure, and vary in length, methods of measurement differ: relative gut lengths are strictly comparable only within studies. Each column represents a separate study.

out of water. It loses water much more slowly than a frog of comparable size. The East African *P. sobrinus*, which spends about 90 per cent of its time out of water, can survive water loss equivalent to a fifth of its body mass (Gordon *et al.* 1978).

The principal nitrogenous waste of aquatic animals is often ammonia, as it is highly soluble and is therefore easily eliminated. Mudskippers produce both ammonia and urea, but while active in air their nitrogen metabolism alters and they produce relatively more urea. It can be inferred that they accumulate both materials while in air, and release them on periodic returns to water.

Respiration in air also requires adaptation. Gills are less effective as a mode of gas exchange in air than in water. It was formerly thought that mudskippers survived in air by taking water into the mouth and gill chambers to keep the gills immersed, and that when the oxygen was depleted or the water lost they had to return to the water to stock up again. It now appears that this is not the case, and that mudskippers are, to a greater or lesser extent, airbreathers.

Periophthalmodon transports air into its burrows and stores it in a dome-like chamber, where it is available as a reservoir of oxygen. This may be particularly important to embryonic development, as mudskippers commonly deposit their eggs on the roofs of their burrows. In several other species, including *Periophthalmus*, *Scartelaos* and *Boleophthalmus*, air is present in the spawning chambers (Ishimatsu *et al.* 1998).

Gas exchange at the gills is supplemented by other parts of the body, particularly the buccal and pharyngeal epithelium and skin. The skin, in particular, is highly vascularised. In *Periophthalmus*, the most important areas for respiration are the front of the head and gill covers. Blood vessels rise from the dermis and branch in the epidermis, forming umbrella-like structures parallel to the surface. Other species have vascularised papillae at the skin surface. The relative importance to respiration of the reduced gills and vascularised skin is reflected in the low values of the gill: skin area ratio for species that are more dependent on aerial respiration. In *Periophthalmus* this ratio is 0.27 to 0.46 and in *Periophthalmodon* 0.22 to 0.5. This compares with a ratio of 0.67 to 0.77 in *Boleophthalmus*, and 0.72 in *Scartelaos*, both species believed to depend less on aerial respiration (Clayton 1993).

5 Measuring and modelling mangroves

Chapters 3 and 4 dealt with the inhabitants of the mangal. The principal characters having been established, this chapter will consider the plot: how do members of the mangrove community interact?

The major transactions between species can be thought of in terms of energy flow or, as energy itself is difficult to measure directly, as carbon fluxes. Nutrient cycles, particularly of phosphorus and nitrogen, are also highly important, and interact with energy flow.

The basis of all ecosystems is autotrophic primary production. Trees are all autotrophic primary producers. Species differ in photosynthetic rate, architecture, salinity tolerance and in many other respects, but these differences are much less significant than the over-riding similarity between species. For practical purposes, mangrove trees can therefore be treated as a homogeneous functional group.

In the same way, sesarmine crabs can be lumped together and treated as a further functional group. Most species subsist largely on decaying mangrove leaves. They are not very different in size, and often remarkably similar in appearance and behaviour. Again, the similarities outweigh the differences. Similarly, deposit feeders, or filter feeders, can be regarded as constituting further functional groups.

This is not to say, of course, that differences between broadly similar species do not matter. To achieve a broad quantitative understanding of a mangrove ecosystem, it is practical to treat, say, sesarmine crabs as a homogeneous functional group and ignore the differences between species, and even the number of species being lumped together. Species diversity within a functional group may well be an important variable: if ten species of sesarmine are present, each with slightly different behaviour and food preferences, is leaf litter processed more rapidly than if the same number of crabs belong to a single species? One major, current controversy in ecology is over the general relationship (if any) between biodiversity and ecosystem function. Are there any important general differences in productivity or stability between a species-rich ecosystem and a species-poor one? The issue is discussed in a later chapter.

By judiciously combining species into functional groups, and analysing their quantitative relationships, it is possible to produce a robust descriptive model of the mangrove ecosystem. Such a model could be refined almost endlessly, but there are limits to the usefulness of such fine tuning. A quantitative model is only as strong as its weakest data, and refining one set of measurements is pointless while another is known only within wide margins.

This chapter attempts to quantify some of the major carbon fluxes through some of the principal functional components of mangrove ecosystems.

How to measure a tree

Photosynthesis and primary production are basic elements of the mangrove ecosystem, as of almost all other ecosystems. Microalgae and bluegreen bacteria contribute significant photosynthesis and primary production on the surface of the mud and of the trees, as well as in mangrove creeks and adjacent waters. The principal primary producers in most mangrove ecosystems, however, are the mangrove trees themselves. Two key features which must be quantified are the biomass of mangrove trees and their rates of photosynthesis and primary production.

Biomass

Biomass is the sum total of all of the components of a tree, below ground as well as above ground: aerial and underground roots, trunk, branches, leaves, flowers and propagules. To be of value, biomass must be related to area. Theoretically, one might simply dig up a given area of mangrove habitat and weigh all of the components of all trees in that area. This heroic operation is scarcely a practical proposition: besides, it would rule out the possibility of recording changes in biomass with time.

Mangrove biomass is therefore usually measured by estimation. The commonest approach is to describe the population structure by measuring individual trees, and to convert this to biomass by using an established relationship between tree size and biomass.

A common practical measure of tree size is the diameter of the trunk at breast height (DBH). An alternative measure is the circumference, or girth, at breast height (GBH): this is generally preferable, since mangrove trunks are often far from circular in cross-section and therefore do not have a single diameter. Because mangrove biologists differ in size, breast height is often taken as a standard 1.3 m. Even then, some flexibility is required in accommodating, say, *Rhizophora* trees where aerial roots spring from the trunk more than 1.3 m above the soil surface, or where a tree divides into two or more separate stems below this height (Fig. 5.1). Further problems arise where mangrove trees are stunted, and a mature tree may never reach a height of 1.3 m (see, for example, p. 195).

Fig. 5.1 A scientist (Liz Ashton: see also Fig. 4.16) attempting to measure the diameter at breast height (DBH) of a *Rhizophora* tree in the Matang forest, Malaysia.

To convert DBH (or GBH) to tree biomass requires cutting down a sample of trees over a range of sizes, weighing them and establishing the relationship between DBH and dry weight or biomass. This relationship is exponential, summarised by the general equation

dry weight $= A \times (DBH)^B$

where A and B are constants.

Alternatively,

$log(dry\ weight) = logA + B.log(DBH)$

It is then possible to arrive at an estimate of the biomass of a tree by measuring its DBH, or of the biomass per unit area by measuring the DBH of all trees within that area.

There are many limitations on the accuracy of this method. The exponential relationship may not hold for the entire range of tree sizes. Seedlings, for example, are very different in shape from mature trees, and are not heavily lignified. It is unlikely that they will fall on exactly the same exponential curve as adult trees. This limitation is probably trivial if an estimate of biomass/area is required, as the collective contribution of seedlings to total biomass is likely to be marginal compared with that of mature trees. A greater problem is that the relationship does not hold for all members of a species under all circumstances, but should be established for each site being studied. Finally, of course, each species will have a different relationship between biomass and DBH.

To move from a measured relationship between biomass and DBH to an estimate of biomass per unit area requires either measurement of every tree in the area, or at least knowledge of the population size structure of a sample of the trees. This can vary widely, particularly after disturbance (Fig. 5.2).

Total above-ground biomass varies widely, typically being highest at low latitudes and declining northward and southwards from the Equator. Some undis-

Fig. 5.2 Size structure (height in metres) of *Avicennia marina* populations at three sites near Xiamen, China. The sample sizes differed and the horizontal bar indicates 10 per cent of each sample.

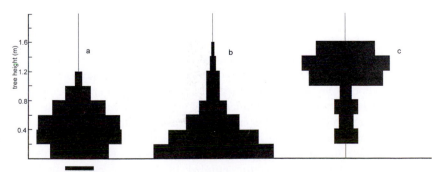

turbed *Rhizophora* forests in northern Australia may reach values of up to 700 tonnes dry weight per hectare (t ha^{-1}), but a range of 500 to 550 t ha^{-1} is more usual in old mangrove forests of South-east Asia, equivalent to perhaps 250 to 275 t carbon ha^{-1}. In contrast, the dwarf *Rhizophora* of Florida may have a biomass of as little as 7.9 t ha^{-1} (Clough 1992; Saenger and Snedaker 1993).

Above-ground biomass consists of leaves, branches, the principal stem of the tree and, in species such as *Rhizophora*, aerial roots. Typically, leaves contribute 3 to 5 per cent of the total, branches 10 to 20 per cent and the main trunk 60 to 90 per cent, while aerial roots, where they are present, may amount to between 8 and 25 per cent (Clough and Scott 1989).

Below-ground biomass is also important, but here there is little reliable information. It is difficult enough to reach a reasonably accurate figure for above-ground biomass, but the problems are compounded with below-ground components. Roots have to be dug out of the mud and cleaned before being weighed. The tangle of absorptive roots must then somehow be separated into live and dead elements. Some reports conclude that mangroves have a high proportion of below-ground biomass compared with other tropical trees: 30 to 50 per cent of the total in *Rhizophora* where much of the root mass is above ground, 50 to 62 per cent in *Avicennia* with its extensive radially spreading underground cable roots (Snedaker *et al.* 1995). In contrast to these estimates, the below-ground biomass of *Rhizophora* in Malaysia has been calculated at less than 20 per cent (Gong and Ong 1990). The ratio of below-ground to above-ground biomass varies with environmental conditions (Chapter 1).

Estimating production

Tree biomass is important when attempting to understand a mangrove ecosystem on a quantitative basis, but it is not itself a measure of production rate, or of energy flow. If biomass is measured in mass per unit area, the logical measure of productivity is dry weight mass per unit area, per unit time. This is commonly expressed as tonnes dry weight per hectare per year, or as grams of carbon per square metre per year. (Carbon amounts to 40 to 45 per cent of tissue dry weight, and 1 gram carbon, in this context, equates to energy of 12 calories, or 50 kilojoules.)

The next step is to estimate primary productivity: the rate at which organic matter is produced as the result of photosynthesis. Direct measurement of photosynthetic rates can be carried out, with suitable instrumentation, in a well-equipped laboratory. In the field, it is generally necessary to resort to indirect and approximate methods. One widely used method of estimating potential primary production entails measuring the attenuation of light as it passes through the mangrove canopy. This can be done simply by measuring incident light directly under the canopy and comparing it with light above the canopy or, in practice, the light reaching the ground in an adjacent open space. If we know the assimilation coefficient of the trees—the number of grams of carbon fixed per unit of chlorophyll—and make some simplifying assumptions about the relative amounts of chlorophyll per leaf in leaves at different levels of the canopy, it is then possible to calculate potential primary production, at least at the moment at which the measurements are taken (Bunt *et al.* 1979; English *et al.* 1997).

This, potentially, estimates **gross primary production**. The advantages of the method are its speed of operation and the simplicity of the field technique. The limitations are that it depends on factors such as the assimilation coefficient which will vary with different conditions of climate and salinity, so at best what is obtained is a 'snapshot' view.

A proportion of the gross primary production is used by the trees in respiration. When these running costs are deducted, what remains is **net primary production**. The rest of the ecosystem depends on this net primary production (and that contributed by other sources such as microalgae).

Respiration rates of leaves can be directly estimated, even in the field, by measuring gas exchange rates. Measuring the respiration rate of the woody components of the tree is less easy, and estimating respiration of the underground root systems is virtually impossible.

Most measures of net primary production have therefore relied on measuring the production of biomass directly. Net primary production should be equal to the increase in standing crop biomass, plus the biomass shed in the form of leaves, twigs, branches and reproductive structures. If net primary production is calculated on the basis of an area of mangrove forest, rather than being related to single trees, the periodic death of whole trees should be included. If a forest is in a state of equilibrium, losses from tree death and shedding of components will balance total biomass production.

Accumulation of biomass is estimated by using the methods of biomass estimation described above, and repeating them after appropriate time intervals, preferably over at least 5 to 10 years. The most reliable estimates come from managed mangrove plantations, where productivity may be very different from virgin forests. In the intensively managed Matang forest of western Malaysia (see Chapter 8) the mean annual increment in above-ground biomass of *Rhizophora* was 18 tonnes per hectare (t ha^{-1}). This is similar to estimated annual increments from Thailand of 14 to 33 t ha^{-1}. An unmanaged section of the Matang, in contrast, gave a much lower estimate of 6 t ha^{-1} y^{-1}. This is low in comparison with unmanaged *Rhizophora* forests in Australia, where accumulation was estimated at 6.3 to 45.4 t ha^{-1} y^{-1} (Clough 1992; Ong *et al.* 1984; Putz and Chan 1986).

In comparison, measurement of litter production is conceptually straightforward, although many practical points need to be taken into account. Appropriate sized mesh litter traps are positioned within the forest in such a way as to trap all material falling from the tree canopy. They must be below the portions of the trees from which material will fall, but high enough to ensure that leaves are not washed away by the tides, or eaten by crabs. The contents must be collected frequently to reduce losses from decay, and over a sufficiently long period to record seasonal and other fluctuations in litter production. Finally, the number of litter traps must be sufficient, not just to achieve statistical validity, but to offset the inevitable losses when local people think of alternative uses for the materials of which they are made.

Despite these practical problems, litter fall measurements have been widely carried out, and often used as a convenient surrogate for full determinations of net primary production (although there is no strong evidence that they necessarily even correlate with each other). Typical figures range from 5 to 15 t ha^{-1} y^{-1}, although in dwarf mangroves, such as those in Florida, production may be as low as 2.9 t ha^{-1} y^{-1}. Litter fall is often strongly seasonal, particularly outside the tropics (Fig. 5.3). (Saenger and Snedaker 1993; Steinke and Ward 1988).

As with biomass, litter production is lower at higher latitudes. The latitudinal trends, however, are not in proportion. The relative amount of litter produced, or litter fall: biomass ratio, increases with latitude. It also decreases with increasing

Fig. 5.3 Mean daily litter production of an *Avicennia* population in South Africa. (Reproduced with permission from Steinke, T.D. and Ward, C.J. (1988). Litter production by mangroves. II. St Lucia and Richards Bay. *South African Journal of Botany*, **54**, 445–54. © South African Association of Botanists.)

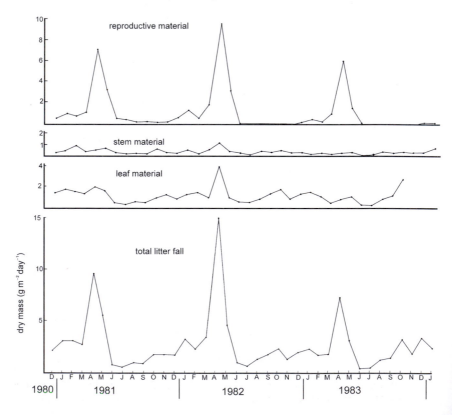

size of tree: small mangrove trees produce relatively more litter than larger ones, presumably because leaves make up a higher proportion of their total above-ground biomass (Clough and Scott 1989; Saenger and Snedaker 1993).

What happens to mangrove production?

As soon as a fallen leaf lands on the forest floor, it is available to other organisms as an energy source. What is the fate of mangrove productivity?

In the short term, the main contribution of the trees is the continual fall of leaves, together with components of flowers and propagules. Although the objective of every propagule is to become a tree, only a tiny minority achieve this ambition (p. 90), and the majority can be considered, together with other components of the litter, as an energy source for heterotrophs.

Branches occasionally drop, and even more infrequently a whole tree will fall. These irregular contributions of woody material must also be considered. The underground components of a mangrove tree also add to the available organic matter. A redundant rootlet is the subterranean equivalent of a shed leaf. In other trees, the never-ending quest for nitrate and phosphate involves production of new rootlets and the abandonment of old ones that have depleted their surroundings of these nutrients. Virtually nothing is known of the turnover of absorptive roots of mangroves, except that it presumably occurs.

Collectively, the dead material shed by mangrove trees is sometimes referred to as necromass (by analogy with biomass). Necromass, of whatever sort, has four possible fates. It may rapidly be broken down, for example by sesarmine crabs (p. 83 and below); it may be decomposed by microbial action; it may be flushed out of the mangal by tides or river flow; or, finally, it may accumulate within the mangrove mud. In most circumstances, very little litter accumulates. The proportion exported will depend on prevailing conditions of tide and current. Export of material may be important to adjoining habitats, and will be discussed in Chapter 6.

The principal routing of organic material within the mangrove ecosystem is through the pathways of microbial decomposition and breakdown by macro-fauna. Figure 5.4 illustrates the immediate fate of litter fall in mangrove forests in tropical Australia, and shows the major differences that depend on shore level. At lower shore levels, tidal movements rapidly flush most of the leaf litter out of the mangrove forest. At higher levels, tidal export is less. If crabs are

Fig. 5.4 (opposite) Initial fate of leaf litter in mangrove forests in tropical Queensland, Australia. Figures in ovals and rectangles represent litterfall and litter standing stock (necromass), respectively, litterfall in g dry weight m^{-2} year^{-1} and standing stock in g dry weight m^{-2}. (Data, with permission, from Robertson, A.I., Alongi, D.M. and Boto K.G. (1992). Food chains and carbon fluxes. In *Tropical mangrove ecosystems.* Coastal and Estuarine Studies no. 41, (ed. Robertson, A.I. and Alongi, D.M.), pp. 293–326, copyright by the American Geophysical Union.)

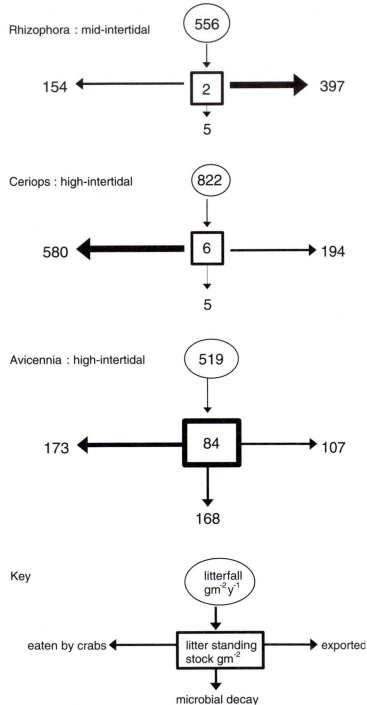

Rhizophora : mid-intertidal

556

154   2   397

5

Ceriops : high-intertidal

822

580   6   194

5

Avicennia : high-intertidal

519

173   84   107

168

Key

litterfall
gm^{-2}y^{-1}

eaten by crabs   litter standing
stock gm^{-2} exported

microbial decay

abundant, they may consume the bulk of the litter; if not, litter tends to accumulate and microbial breakdown is more important.

Microbial breakdown

Left to decay, leaves decline in mass. Decay rates can most readily be assessed by placing known masses of mangrove leaves in mesh litter bags which exclude macrofaunal leaf eaters such as gastropods and crabs, leaving them in various positions within the mangal and weighing the remains at suitable intervals. As an example, Fig. 5.5 illustrates the rate of decline in dry weight of *Sonneratia* leaves at a mangrove site in western Malaysia.

Decomposition is more rapid in leaves immersed in mangrove creeks or adjacent waterways, or which lie in a position frequently inundated by the tides. Because the decline in weight is approximately exponential, decomposition rate can conveniently be summarised as the time taken for mass to fall to half of its original value, or half-life. The half-life of *Avicennia* leaves which are permanently immersed is 11 to 20 days, while in the intertidal zone it may be 90 days. Rate also varies with species: *Rhizophora* leaves, for instance, break down at about half of the rate of *Avicennia* (Robertson *et al.* 1992).

For the first 10 to 14 days, almost all of the loss in weight is due to physical leaching of dissolved organic carbon (DOC: sometimes referred to as dissolved organic matter, DOM) from the leaves, rather than by microbial decay. Approximately 30 to 50 per cent of the organic matter of the leaves is leachable in this way, and much of what remains is insoluble structural carbohydrate such as cellulose. This is subsequently attacked by extracellular enzymes secreted by bacteria or fungi. Among the substances which leach out over the first few days are tannins (p. 87), which inhibit bacterial growth, probably by precipitating secreted enzymes and soluble substrates. Presumably

Fig. 5.5 Decline in dry weight with time of *Sonneratia* leaves in the Matang, western Malaysia, as percentage of initial dry weight. (Data: Liz Ashton.)

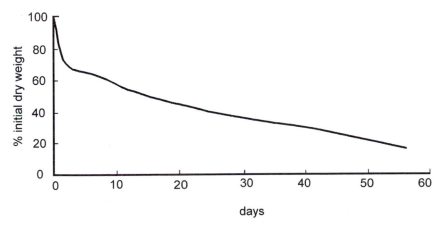

the bacterial invasion takes place after tannin levels have fallen (González-Farias and Mee 1988).

After a few days, the leaf surface has been colonised by a dense bacterial population; bacterial densities of up to 600 000 000 cells per square centimetre of leaf surface have been recorded after only 6 days, with a production rate of up to 8 000 000 bacterial cells per square centimetre per hour (Benner *et al.* 1988). Numerous species of bacteria are probably involved, although very little is known about their differing roles and interactions, or about the ecology of mangrove bacteria in general. One study of bacterial decomposition failed to isolate any bacteria capable of digesting cellulose, so it may be that bacteria are more important in processing the soluble matter that leaches out. The leached material tends to form an aggregate, which, after colonisation by microbes, develops a carbon: nitrogen (C: N) ratio of around 6 (see Fig. 4.5). It is therefore a highly nutritious food source for small invertebrates, and is one of the main constituents of particulate organic matter (POM; alternatively particulate organic carbon, POC) in the sediment (Robertson *et al.* 1992; Steinke *et al.* 1990). (The distinction between particulate and dissolved organic matter (POM and DOM) is to some extent an artificial one, since in fact there is a spectrum from truly dissolved material, through colloidal suspensions, to sizeable and discrete particles.)

Slightly later in the succession, fungi in large numbers appear on the decomposing leaf surface. The role of fungi in mangrove leaf decomposition is also poorly understood. Many species are present, some occupying leaves soon after they reach the mud surface, others only after some time. Two genera in particular, *Trichoderma* and *Fusarium*, are powerful digesters of cellulose and play a major part in leaf breakdown (Hyde and Lee 1995; Steinke *et al.* 1990).

Bacteria and fungi actively break down leaf material, and convert much of it into their own biomass. The conversion efficiency is 40 to 70 per cent for fungi, and around 50 per cent for bacteria (Steinke *et al.* 1990). Microbes themselves constitute a major food source for other organisms, such as deposit-feeding fiddler crabs. Even the sesarmines which overtly feed on actual leaf material may in fact be extracting much of their nourishment from the colonising microbes (p. 87). Meiofauna (p. 110) are also found associated with decaying leaves, and many feed on microbial cells. Like microbes, they are inhibited by tannins (Alongi 1987*b*).

Crabs and snails

Crabs and snails usually consume a significant proportion of leaf litter. How important is this to energy flow through the ecosystem?

As with trees, crab productivity—in this case secondary, not primary, productivity—must be measured. The problem is essentially similar to the measurement of net primary production of trees, where some ingenuity is needed to assess biomass increase and litter fall. The biomass of sesarmine crabs per unit area is analogous to the standing crop biomass of trees. Crabs are distinctly more mobile than trees, and can hide down burrows: this makes biomass estimation difficult, but not impossible. The parallel to leaf litter fall is the combined mass of crab faeces, moultskins and larval release. This is measured, not by litter traps, but by estimates based on laboratory observations of faecal production and moulting frequency, and knowledge of egg production rates and larval weights. Crab mortality to predators must also be assessed, since if the crab population is roughly in equilibrium this harvest will account for the major part of crab production.

Given the difficulties, it is not surprising that there have been few estimates of sesarmine crab productivity. On one Malaysian mangrove shore, the two major species of sesarmine, *Sesarma (Perisesarma) onychophorum* and *Episesarma versicolor*, were present at average densities of 4 and 0.3 individuals per square metre, corresponding to biomasses of 2.56 and 1.95 grams per square metre (g m^2). Annual production was estimated at 0.7 and 9.1 g m^2, respectively (Macintosh 1984).

Leaf-eating crabs are important for energy and carbon flow through the mangrove ecosystem in several distinct ways. By converting leaf into crab material they provide a food source for predators such as fish and bird—and, of course, other crabs. They influence energy flow in two ways other than by assimilating leaf material directly. In the first place, as described in the last chapter, they remove a high proportion of the leaf litter and conceal it in burrows. Whether or not it is then eaten, the leaf material is retained within the mangal and not flushed out by the tide or currents.

Finally, and possibly most importantly, a high proportion of leaf material is returned to the environment in the form of faeces. Crab faeces are richer in nitrogen than uneaten leaf material: passage through a crab gut enhances the nutritional quality of mangrove leaves (Lee 1997). The hindgut contents of a sesarmine often consist almost entirely of small leaf fragments, often similar in appearance to those in the foregut. A key role of the crabs is to break down the tough leaf material into smaller fragments. This will accelerate further break-down by increasing the surface area available for microbial action in the mangrove sediment, and will also produce smaller particles suitable for other, deposit-feeding, animals. In one Australian *Bruguiera* forest, sesarmine crabs produce faeces equivalent to around 260 g carbon m^2 y^{-1} (Robertson and Daniel 1989). Overall, crabs accelerate litter breakdown by up to two orders of magnitude (Robertson *et al.* 1992). From an ecosystem perspective, sesarmines are more important as food processors than as assimilators.

Wood

Woody tissue makes up 20 to 50 per cent of the total net primary production of mangrove forests, so its fate is important to understanding energy and carbon flow through the ecosystem. Wood is also broken down, although much more slowly than leaf litter. Lignocellulose, which makes up much of the bulk of woody material, is highly refractory to decay. Trunks and large branches disintegrate over a timescale of a few years, compared with weeks or months for leaves. Small twigs break down more rapidly.

The initial breakdown of wood is largely accomplished by animal activities. If a mangrove tree dies standing, termite activity rapidly removes much of the woody material. Fallen branches and trees become honeycombed by burrows of teredinid molluscs (shipworms: p. 109), aided by cellulose-digesting bacterial symbionts. In tropical Australia, it has been observed that within 4 years about 40 per cent of the cross-sectional area of a fallen *Rhizophora* trunk has been replaced by shipworm tubes, accounting for about 90 per cent of the loss in wood mass over that period. Half of the original mass is lost within 2 years where teredinids are present, and only about 5 per cent where they are absent (Robertson 1991). Along with shipworm burrowing, microbial decay of the remaining material takes place, particularly through the actions of fungi.

The only studies of root decomposition, on temperate *Avicennia*, indicate that fibrous roots decay more slowly than the major radial roots. In 270 days, fibrous roots lost only 15 per cent of their mass, compared with 60 per cent for the major roots (van der Valk and Attiwill 1984). This probably reflects the enormous surface area within the major roots, on which bacteria operate, which results from the air spaces within the aerenchyma (p. 5).

The process of breakdown of wood litter is slow, but may contribute as much energy and carbon to the ecosystem as leaf litter. If woody debris is not building up, then the rate of breakdown matches the rate of woody litter production. The rate at which breakdown products are made available to other organisms in the ecosystem matches the rate of litter production, even if the intervening processes have taken a long time.

Role of sediment bacteria

Bacteria are important in facilitating the breakdown of mangrove litter. In addition to this role, sediment bacteria are an important element in the flow of carbon through the mangal as a whole. In the top 2 cm of mangrove sediment there may be up to 3.6×10^{11} bacterial cells per gram (dry weight) of sediment, and productivity of 5.1 g carbon m^{-2} d^{-1}, equivalent to an annual production of more than 18 t ha^{-1}. These are among the highest values known for any marine sediment (Alongi 1990).

Sediment bacteria are an important food source for some meiofaunal species, as well as for deposit-feeding macrofauna. In general, however, meiofauna are not a major element in limiting bacterial populations, or in determining carbon flux rates.

Most bacteria die uneaten, either because they locally exhaust their substrate ('local' to a bacterium being at a very small scale indeed) or, perhaps by the action of bacteriophage viruses. When bacteria die, soluble materials are released and add to the pool of dissolved organic carbon (DOC) available for further bacterial activity. Little of this escapes from the sediment, unless all bacterial activity is stopped, for instance by poisoning with mercury salts. DOC is in turn acted on and assimilated by bacteria, and so becomes particulate again. In this way, cycling of carbon may take place entirely within the bacterial community, which in effect constitutes a 'carbon sink' within the mangal. In Malaysian mangroves it has been estimated that, overall, some 1.5 t carbon ha^{-1} have been accumulating in the sediment each year, about 10 per cent of total production (Ong 1993; Robertson *et al.* 1992).

The fate of organic particles

Digested by leaf-eating crabs and voided as faeces, or broken down by microbes, much of the mangrove productivity ends up sooner or later as small organic particles (POM). As with the original leaf litter, some of this is washed out of the mangal, but the bulk remains. Together with bacterial cells, diatoms and other matter, it constitutes the food source for fiddler crabs and other deposit feeders, and filter feeders such as oysters.

Fairly obviously, there is a continuum of particle size, from whole leaves (or even fallen trees) to the smallest fleck of organic matter in the mud, or suspended in the water. Figure 5.6 shows that there may also be a continuum of feeders on particles of detritus, each specialising in a different range of particle size.

Of all of the mangrove deposit and filter feeders, the fiddler crabs (*Uca*) are probably the best known. Chapter 4 discussed their population density, sediment processing rates and productivity. Clearly fiddlers play a significant role in cycling organic matter within the mangal, by ingesting particles and returning

Fig. 5.6 Mean particle size in the stomach or buccal contents of animals in a Florida mangal. (Reproduced with permission from Odum, W.E. and Heald, E.J. (1975). The detritus-based food web of an estuarine mangrove community. In *Estuarine research*, Vol. 1, (ed. L.E. Cronin), pp. 265–286, © Academic Press.)

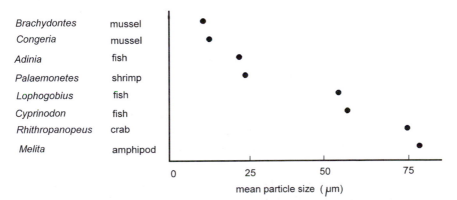

faeces to the soil surface. They also contribute to export, through their massive production of larvae, as well as by being eaten by commuting predators. Other, less charismatic, deposit and filter feeders are comparatively neglected, but undoubtedly have similarly important roles.

Predators

Relatively few macrofaunal predators are permanently resident within the intertidal zone of the mangal: the majority, such as birds, fish and the larger swimming crabs, are highly mobile and commute to and from either tidally, seasonally or both. Largely for this reason, there is virtually no quantitative information of feeding rates or productivity. Even where such information is available, it is hard to estimate how much of the productivity should be attributed to mangrove sources and how much to feeding elsewhere. The rich harvests of crabs, shrimp and fish from waters close to mangroves indicate the general importance of mangrove productivity to these animals (see Chapter 6).

Crocodiles, alligators and caimans are among the most striking of top predators in mangrove areas. In the Sundarbans of India and Bangladesh, the estuarine crocodile (*Crocodylus porosus*) is an important predator of fish (Chapter 3). It might be expected that these voracious predators would have a distinctly negative effect on fish populations, hence on human fisheries. In fact the effect is opposite to that expected: as it happens, crocodiles tend to eat the less commercial fish, some of which in turn feed on smaller fish, and their eggs, of species which are more sought after by humans. Crocodiles, in fact, benefit human fisheries by reducing predators of commercial species. In addition, by their movements they enhance the movement of nutrients in the water, helping to maintain productivity. A similar effect has been noted with caimans (*Caiman crocodylus*) in the Amazon. It was noted that wherever caimans disappeared, fishing subsequently declined. In this case nutrients released by caimans in the course of their metabolism, mostly originating from outside the area, were sufficient to have a stimulating effect on primary production (FAO 1994).

These examples give some idea of the complexity of the interactions of predators and prey, and of the possible importance of such interactions, in mangrove ecosystems.

Putting the model together

If the major functional components of the mangrove ecosystem, and the fluxes of carbon between them, have been measured, it is possible to assemble the information into a food web, or model of the entire ecosystem. Figure 5.7 summarises the major interactions between the elements of a mangrove ecosystem in north-eastern Australia. Other ecosystems, of course, would show differences in the emphasis on the different components, and it is clear that certain of the components have not been adequately measured.

Fig. 5.7 Major carbon fluxes through a mangrove ecosystem. NPP = net primary production, DOC = dissolved organic carbon, POC = particulate organic carbon. Masses in tonnes carbon per hectare (tC ha^{-1}), fluxes in tonnes carbon per hectare per year (tC ha^{-1} ha^{-1}). Below-ground biomass and fluxes omitted. (Mainly after Robertson, A.I., Alongi, D.M. and Boto K.G. (1992). Food chains and carbon fluxes. In *Tropical mangrove ecosystems*. Coastal and Estuarine Studies no. 41, (ed. Robertson, A.I. and Alongi, D.M.), pp. 293–326, copyright by the American Geophysical Union.)

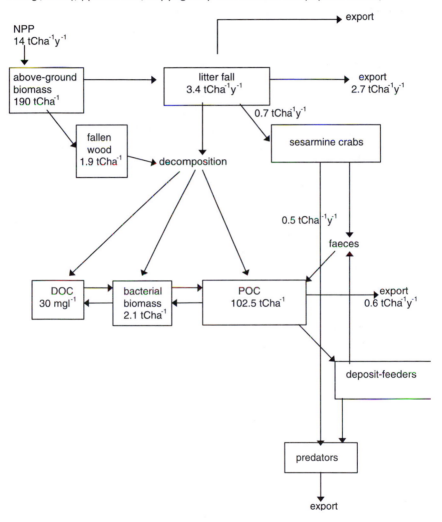

The example shown is a very much oversimplified one. A number of major components have been omitted, only very broadly-defined functional groups of organisms are considered, and the model includes only carbon masses and fluxes. A similar diagram could be produced to include the movements and accumulations of nitrogen, phosphorus or any of the other key nutrients: or, ideally, of all of these together. As it stands, the diagram is perhaps quite complicated enough to give a general idea of the approach.

Why attempt to model ecosystems? In the first place, a quantitative understanding is more satisfying than a qualitative one, and not just as an exercise in accountancy. It informs us, for example, not just that sesarmine crabs as a group eat mangrove leaves, but that that they eat a significant proportion of leaf litter and are therefore crucial to carbon (and energy) flow through the ecosystem. In other mangals, such as those of Florida, this would not be the case, so illuminating comparisons can be made between different areas. Secondly, even a simple model can make gaps in knowledge and understanding obvious, and clarify priorities for further research. If a model converges sufficiently closely with reality, it can virtually be used as an experimental system, or at least to suggest fruitful experiments that could be carried out. Finally, a sufficiently detailed model can be used as a powerful planning and conservation tool, in predicting the probable effects of future impacts on a mangrove ecosystem.

Perhaps the best example is the classical analogue model compiled by Lugo *et al.* (1976), based on the mangroves of southern Florida, and designed to run as a computer simulation. This incorporated not just biomass measurements and carbon fluxes, but solar irradiation, photosynthetic and respiration rates, tidal activity, nutrient availability from various sources and its effect on photosynthesis and other factors.

Some interesting predictions emerged when this model mangal was allowed to run for a simulated period of many years under various settings of the key environmental variables. If it was started at a simulated early stage of succession, it reached a steady state of biomass in 23 simulated years, at a rather higher biomass (per unit area) than that measured at a real site in the vicinity. Hurricanes occur in the area with a mean frequency of around 20 years, so real mangrove stands will, for the most part, be composed of trees younger than 20 years, with biomass/area levels lower than the maximum potential. The model therefore, suggests a possible explanation for the observed values.

Running the model under different conditions of nutrient availability suggested that for characteristic growth rates to be maintained, a steady input of terrestrial nutrients was required, and that nutrient availability rather than salinity was the major limiting factor on mangrove growth. This gave a valuable (and directly testable) insight into mangrove ecosystem function. The information was also directly relevant to management of the Florida mangroves. Freshwater runoff from the land, carrying dissolved nutrient, has been progressively canalised and diverted so as to bypass the mangroves and run directly into the sea. The prediction derived from the simulation model is that this will cause further decline in mangrove growth and productivity.

6 Comparisons and connections

A classical mangrove ecosystem is a soft-bottomed intertidal community, highly productive and with a diverse fauna. It is dominated by angiosperm trees which are adapted to a saline, waterlogged and anoxic substrate. The emphasis in energy flow through the ecosystem is on detrital breakdown by a combination of detritivores and microbial decomposition pathways, the former complementing and facilitating the latter. In general, control is bottom-up, with primary production regulated by nutrient availability and salinity, secondary producers by primary producers and so on. Finally, and for reasons that are not clear, mangroves are almost exclusively tropical.

With the exception of the mangrove trees themselves, this description could apply to a range of tropical, soft-bottom, coastal communities. How different is a mangrove habitat from its immediate surroundings? Is a mangal just a stretch of intertidal mudflat with trees, and a consequential concentration of fauna? Apart from the presence of trees, are the differences between mangal and mudflat communities qualitative as well as quantitative?

Even if mangrove habitats are distinctively different from their surroundings, they are not isolated from those surroundings. Interchanges take place between the mangal and its immediate vicinity, of carbon and inorganic nutrients, of eggs and larvae, of commuting invertebrates, fish, mammals, reptiles and birds. An understanding of these connections places mangrove habitats in their eco-logical context.

Is the mangrove ecosystem unique? This chapter finally considers the simila-rities and differences between mangrove ecosystems and their equivalent in temperate latitudes, salt marshes.

How distinctive is the mangrove community?

Mangrove trees themselves are distinctive. The principal species are found in mangrove habitats and nowhere else, and clearly define the limits of the mangrove habitat. Within this habitat live many other organisms, collectively constituting the mangrove community. These are discussed in Chapters 3 and 4. A key question is whether the species present are so specialised that they cannot live elsewhere, or whether they are drawn from the surrounding pool of species, concentrated in the mangal merely because it offers richer resources than those available in the immediate vicinity.

Relatively few species are mangrove-specific. Some instances were discussed in previous chapters. Among the vertebrates, only a few birds and fewer mammals and reptiles are restricted to mangroves, while fish forage among the trees only at high tide. Many grapsid and ocypodid crabs (p. 81) are regarded as characteristic inhabitants of mangroves, but species actually known to be more or less exclusively found there are surprisingly few: four or five have been found only in mangroves, with a further dozen or so possible candidates.

The evidence is not very reliable, partly because biologists working in mangrove areas tend to pay less attention to adjoining mudflats; and, of course, because there is little interest in recording a species' absence from a habitat. One of the few studies focused directly on the question of mangrove faunal specificity is a survey of coastal marine habitats of the entire Saudi coast of the Red Sea. This compared the species composition of seagrass and other littoral and shallow-water habitats, as well as mangroves, and clustered the data from collecting sites by faunal similarity. Mangrove habitats showed no greater overall similarity to each other than they did to other soft-bottom habitats.

Red Sea mangroves are not typical of mangroves world-wide. Tree species diversity is low (predominantly a single species, *Avicennia marina*), the substrate is usually sand rather than mud with little accumulation of leaf litter, mangrove stands are coastal fringing rather than estuarine and most are small in extent. All of these factors are likely to contribute to the low diversity of mangrove invertebrates and to make the presence of a specifically mangrove fauna less likely. However, typical or otherwise, here at least there is no evidence of a characteristic mangrove fauna. Species are shared with surrounding habitats (Price *et al.* 1987).

The majority of mangal animals may therefore be opportunists, facultative rather than obligate inhabitants of mangroves. The mangrove community is to a large extent a concentration of species drawn to the mangrove trees, and flourishing there, because of the abundance of detritus and the physical habitat offered. If species are, in general, not restricted exclusively to mangrove habitats, individuals are likely to move to and fro between mangroves and their surroundings. If this likely interchange of organisms is taken together with the probable interchange of nutrients and organic matter, it is obvious that mangrove ecosystems cannot be considered in isolation from their surroundings.

Exchange with the surroundings

To what extent do mangroves interchange materials with their surroundings? Clearly, this will vary considerably, depending on the environmental setting of the mangal concerned. At one extreme, the rare mangroves which grow in complete isolation from the sea (p. 35) are unlikely to export much of their productivity to surrounding habitats. The most significant means of export would be by animals feeding within the mangal and subsequently moving elsewhere. At the opposite extreme, a mangrove flushed by a large river, or by frequent tidal inundation and active currents, is likely to lose a much greater proportion of its litter production. Rivers and tides can also bring in nutrients and particulate matter. Soluble nutrients (particularly nitrates and phosphates) can then be retained and used by the mangrove trees, and organic particles trapped in the mangal (p. 47), so that imports as well as exports may be relatively high.

Such physical imports and exports of material are discussed in the next section. Materials are also imported and exported actively, in the form of animal movements in and out of the mangal. These can represent considerable transfers of material. The amount of inorganic nitrogen and phosphate deposited by 220 000 roosting fruit bats (p. 75), or the annual removal of carbon and nitrogen represented by a flourishing fishery, are not negligible. These could perhaps be estimated with reasonable accuracy. It is more difficult to assess the influence of transient visitors such as fish or shrimp commuters. The relations between fish, shrimps and mangroves are discussed later.

Outwelling

Early studies of salt marshes indicated that they were highly productive ecosystems, and that much of this production was probably exported, where it sustained secondary consumers and food chains in nearshore waters and might contribute significantly to commercially important fisheries. The general proposition, that productive coastal ecosystems were net exporters of organic matter to adjacent waters, was soon extended to mangroves. This 'outwelling' effect was seen as an important feature of mangroves, and a strong argument for their conservation.

The outwelling hypothesis rests initially on a fairly simplistic balancing of the ecological books. Primary production can be measured, partially in terms of litter fall, or more fully by the methods discussed in the last chapter. Consumption by herbivores, and decomposition within the mangal, can also be measured, although probably with less reliability. If production is greater than consumption and *in situ* decomposition, then the difference can be assumed to represent export from the mangal. The estimate can be made more realistic if net accumulation of matter within the mangal is considered. Imprecision in the original measurements is compounded in the final figure but, failing any fuller data, this mass balance approach can give a reasonable provisional indication of the movement of organic matter out of the mangroves.

This approach inferred export indirectly from apparent shortfalls between estimates of production and of fate. Export can also be estimated directly, by combining information on the movements of water through the mangal with measurements of its content of particulate and dissolved organic carbon.

Inorganic carbon in soluble form may also be significant, but has virtually never been considered.

Mangrove habitats are often topologically complex, and water movements of course vary with currents, tides and meteorological conditions. To give a complete picture, measurements have to be carried out over a prolonged period of time, to include seasonal and other variation. A brief rainy season may flush out leaf litter or inorganic substances which have been steadily accumulating, resulting in a pulse of exported nutrients to the surroundings which would be missed in a short-term study. Merely collecting the raw data, before assembling it into a full model of carbon movement, is therefore a task of Herculean proportions. Not surprisingly, few such comprehensive studies have been achieved.

As usual, one of the most thorough analyses comes from the mangroves of Hinchinbrook Island, Queensland, in this case a study of the tidal mangrove forests on Coral Creek, Missionary Bay. A proportion of the litter fall was sequestered by crabs: after allowing for this, it was estimated that about 7.5 kilogram carbon per hectare per day (kg C ha^{-1} d^{-1}) was exported as recognisable leaf litter, and a further 1.6 kg C ha^{-1} d^{-1} in the form of particulate organic carbon (POC). Together, these components comprised around 30 per cent of leaf litter production. Dissolved organic carbon (DOC) was also measured, but DOC exchange was trivial in comparison, amounting to a net import of only 0.2 kg C ha^{-1} d^{-1}. Labile forms of dissolved organic carbon appear to be efficiently recycled within the forest, rather than exported (Boto *et al.* 1991; Robertson *et al.* 1992).

A contrast to the productive mangroves of Australia is shown by high intertidal forests in Florida, where the total export of particulate carbon was only 0.4 kg C ha^{-1} d^{-1}. Here DOC was, relatively, of more significance, amounting to an export of 1.0 kg C ha^{-1} d^{-1}, more than 70 per cent of total carbon exported (Twilley 1985).

As already discussed, the fate of primary production varies greatly, depending in particular on hydrological conditions. Riverine mangroves may acquire an input of POC and DOC with the river water, as well as contributing their own productivity to the outflow, and a tidally-dominated mangal is likely to export more of its production than a high shore, or even inland one (see Fig. 5.4). Except in a very few cases, there is simply insufficient information to establish the scale and significance of organic carbon outwelling; the only valid generalisation is that generalisation is impossible.

The position with respect to other materials, particularly nitrogen and phosphorus, is even less clear. Again, Missionary Bay provides the most thoroughly-investigated example. The principal sources of nitrogen are the products of nitrogen fixation within the mangal, and dissolved organic nitrogen, nitrate, nitrite and ammonium brought in by tidal flow. Rapid recycling of nitrogen takes place within the mangal (see Chapter 1), and of course large amounts of nitrogen are static there in the form of standing crop biomass—perhaps 6000 t of nitrogen in above-ground biomass alone. Both soluble and particulate nitrogen are exported. Overall, the balance is for a considerable net import of soluble forms of nitrogen, and a net export of particulate nitrogen (Table 6.1). As far as available information goes (which is not very far), this seems fairly typical of mangroves. The phosphorus budget is less clear, but there appears to be a significant net annual import, corresponding to nearly a quarter of what is required for the known rates of primary production (Alongi *et al.* 1992; Boto *et al.* 1991).

The fate of mangrove exports

Given the extent of carbon export from mangrove ecosystems, marked effects on neighbouring ecosystems would be expected. What is the fate of the out-welled carbon, both particulate and dissolved, and what are its effects?

Physically identifiable mangrove detritus does not generally go far. In one study of mangroves in southern Florida, for instance, the percentage of mangrove material in plant litter fell off from 80 per cent to about 10 per cent within less than 100 m of the edge of the mangrove forest (Fig. 6.1) (Fleming *et al.* 1990). Although dispersal will depend on local conditions of river flow, tides and currents, leaf litter probably has only very local effects.

Much of the exported organic material is in the form of small organic particles, or dissolved matter, which are not so readily identifiable. Dissolved organic matter from mangroves (DOM) can be identified by the analysis of lignin-derived substances, and naturally fluorescing compounds which could have come only from vascular plants. On this basis, about 10 per cent of the DOM detected in the coastal waters of the Bahamas appeared to have come from mangroves at least 1 km away. Correlation between the concentration of mangrove-derived DOM and bacterial production suggested that this was significantly contributing to bacterial production in the plankton (Moran *et al.* 1991).

Measurement of the ratio of stable isotopes of carbon has been widely used as a technique for tracking carbon through food chains. The method was discussed in Chapter 4. The ratio, relative to a standard, is expressed as a $\delta^{13}C$ value, in

Table 6.1 Nitrogen budget from Missionary Bay mangrove forest, Queensland, data modified from Alongi *et al.* (1992)

Process	Input	Output	Balance
nitrogen fixation	36 831		36 831
denitrification		2824	−2824
tidal: dissolved	168 598	116 108	52 490
tidal: particulate		76 321	−76 321
total	205 487	195 253	10 234

Figures are in kg nitrogen per year. Negative figures in the balance column indicate net export.

Fig. 6.1 Percentage of identifiable mangrove detritus along a transect outwards from a riverine mangrove forest into a sea grass bed in southern Florida. The horizontal bar indicates the extent of the mangrove forest. (Reproduced with permission from Fleming *et al.* (1990). Influence of mangrove detritus in an estuarine ecosystem. *Bulletin of Marine Science*, **47**, 663–9.)

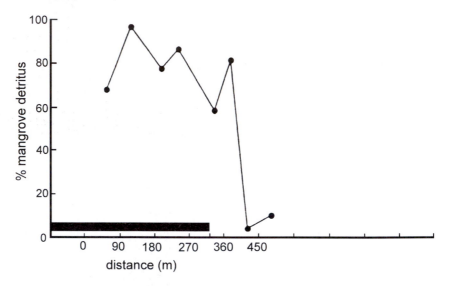

parts per thousand. Mangrove leaves have $\delta^{13}C$ values in the range $-24‰$ to $-30‰$, while other marine plants have a higher (less negative) $\delta^{13}C$ values. The isotope composition of a consumer reflects that of its diet, with only a slight shift due to selective handling of the two isotopes during respiration.

Figure 6.2 shows the range of $\delta^{13}C$ values in organic matter from sediments, and in the tissues of animals collected from mangroves and their associated mudflats in western Malaysia, from coastal inlets within 2 km of the mangroves, and in offshore waters between 2 and 18 km away. Even animals collected from within the mangal itself have $\delta^{13}C$ values, on the whole, which differ from the range found in mangrove material. This shows that their primary carbon source is not mangal material. Any influence of mangrove carbon wanes still further in nearshore waters, where the principal carbon source, on this evidence, seems likely to be phytoplankton, while offshore animals seem on this basis to have no dependence whatever on mangrove-derived carbon (Rodelli *et al.* 1984).

Fig. 6.2 (opposite) Distribution of $\delta^{13}C$ values (‰) in the tissues of animals caught within a Malaysian mangrove swamp, in coastal inlets among the mangroves, and offshore. The horizontal bars below indicate the range of $\delta^{13}C$ values from mangrove leaves and algal material. Values of sediment material were $-24.8 \pm 0.9‰$ (within mangrove forest), $-26.2‰$ (mudflat) and $-24.8‰$ (56 km offshore). (From data in Rodelli *et al.* (1984). Stable isotope ratio as a tracer of mangrove carbon in Malaysian ecosystems. *Oecologia*, **61**, 326–33, © Springer-Verlag, reproduced with permission.)

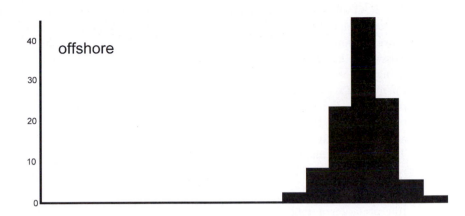

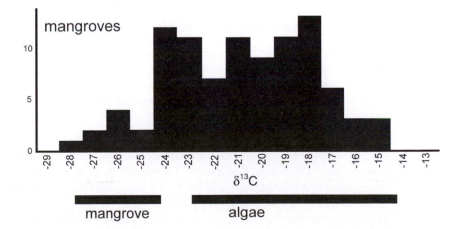

The $\delta^{13}C$ values of carbon in nearshore and offshore sediment samples is closer to that of mangrove material, and it appears that a large proportion of mangrove carbon is initially deposited in sediments. Returning to the studies on Missionary Bay, isotope studies were used to estimate the fate and disposition of mangrove carbon after it was exported from the mangal. Sites 1 to 2 km from the mangal accumulated at the rate of around 440 kilogram carbon per hectare per year (kg C ha^{-1} year^{-1}), and even at sites 10 km from shore the accumulation rate was 10 to 40 kg C ha^{-1} year^{-1}. This accumulation of material appeared to have little stimulating effect on bacterial or meiofaunal activity. The mangrove material appears to be either refractory, possibly because of high residual tannin levels, or simply of little nutritional value until colonised by bacteria (Robertson *et al.* 1992; Torgerson and Chivas 1985).

Interpretation of stable isotope ratios is not always straightforward. The general inference that consumers collected within the mangal do not depend directly on mangrove carbon depends on which species are collected. Values of $\delta^{13}C$ higher than those of actual mangrove leaf material, but lower that those of alternative primary producers suggests that the species sampled utilised more than one carbon source in their diet, rather than that they do not use mangrove-derived carbon. Finally, tissue $\delta^{13}C$ values depend on the number of intervening stages in the food chain as well as on the source of primary production.

With these reservations, however, it can be deduced that much of the carbon exported from mangroves (at least from the few sites studied to date) settles initially in sediments within, at most, a few kilometres of the source. Here it is processed, perhaps slowly, by bacterial action. Possibly successive cycling through bacterial metabolism erases the $\delta^{13}C$ signature characteristic of mangrove material. Eventually much of the carbon is released into the water column as DOC and mixed with carbon from other sources, and in this form it may re-enter planktonic and macrofaunal food chains, either in the same location or dispersed over a much wider area.

Mangroves, seagrasses and coral reefs

Seagrass beds often grow in close proximity to mangroves, and the two ecosystems are often closely linked through fluxes of carbon and other materials. Seagrasses are, like mangroves, vascular plants. They often cover vast areas, in shallow soft-bottomed coastal areas throughout the tropics and in temperate seas.

In many respects, seagrass beds provide interesting parallels with mangrove forests, besides the basic similarity of being habitats dominated by salt-tolerant vascular plants. The number of genera and species in each category is similar. Both mangroves and seagrasses have adapted physiologically to the difficulties of living in marine habitats. Both are a major source of primary productivity in areas where alternative primary producers are not particularly abundant, and both help to consolidate soft substrates in what would otherwise be unstable, soft sediments. Both also provide a firm physical base for attachment by other organisms, and are occupied by more or less characteristic faunal assemblages, although with few exclusive species. Finally, much of the productivity of both mangroves and seagrasses enters food chains through detrital pathways rather than by direct herbivory of the growing plant (Phillips and McRoy 1980).

Links between mangroves and seagrasses have again been elucidated by stable isotope studies. Fortunately, seagrass show $\delta^{13}C$ values much higher (less negative) than those of mangroves, making it possible to identify an influence of mangrove carbon within a seagrass bed. In southern Florida, fronds of the seagrass *Thalassia* growing close to mangroves had more negative (that is, more mangrove-like) $\delta^{13}C$ values than those collected far from mangroves. The same was true of carbon in calcium carbonate of mollusc shells collected from sites near to, and far from, the mangroves (Table 6.2). This strongly suggests a short-range influence of mangrove carbon, probably through the assimilation by the seagrasses (and subsequently molluscs) of CO_2 released by the processing of mangrove material (Lin *et al.* 1991).

Similar results from a mangrove forest and adjacent seagrass beds in Kenya suggested a similar conclusion. Here there was a gradient in $\delta^{13}C$ values from the mangroves through the beds of seagrass (*Thalassodendron*), both in the sediments and in seagrass leaves. The most probable interpretation is that particulate organic matter (POM) from the mangroves is distributed in the sediments in amounts which decrease rapidly with distance from the source, and that microbial respiration in the sediment then makes CO_2 from the POM available for seagrass photosynthesis. During flood tides, POM appears to flow back from the seagrass beds into the mangal. The two ecosystems are thus tightly coupled by reciprocal movements of carbon (Hemminga *et al.* 1994).

Mangroves and seagrass beds are also connected by the movement between them of actively-swimming animals, particularly crustaceans and fish. These either commute between the habitats, or occupy them at different stages of their life cycles (see p. 142).

Connections between mangroves and coral reefs are more elusive, but may be important. Corals are very vulnerable to sedimentation and to eutrophication which results from nutrient enrichment. In some situations, therefore, mangroves may protect reefs by trapping river sediments, and sequestering

Table 6.2 $\delta^{13}C$ values, in parts per thousand (‰) ±s.e., for seagrass leaves and mollusc carbonate collected close to, and far from, mangrove stands, from Lin *et al.* (1991)

	Near	Far
Seagrass leaves	-12.8 ± 1.1	-8.3 ± 0.9
Mollusc shell	-2.3 ± 0.6	0.6 ± 0.3

The $\delta^{13}C$ value for mangrove carbon normally falls in the range −24‰ to −30‰.

nutrients. Mangroves, seagrasses and reefs all tend to export organic matter, both dissolved and particulate (DOM and POM). There appears to be a net movement of DOM in the direction: mangroves → seagrasses → coral reefs. The situation with POM is more complex, and neither DOM nor POM movement has been adequately quantified. Energy also flows between these habitats in the form of animal movements. Some reef fish may be among those that use mangrove habitats as nursery areas, and fish and invertebrates move between reefs and mangroves, as between seagrass beds and mangroves. There are therefore significant linkages between the three types of habitat, although their quantitative importance has not adequately been assessed (Ogden and Gladfelter 1983).

Commuters

Passive transport of dissolved and particulate organic matter is not the only connection between mangrove and other habitats. Considerable amounts of organic matter move to and fro actively, particularly in the form of fish and crustaceans, and although these movements are difficult to assess quantitatively in terms of carbon flux, they are nonetheless important.

There are different ways in which a mangrove habitat can be exploited by a species which spends only part of its time in that habitat. Individuals might commute in periodically to feed, or to use the mangrove environment as a refuge from predators, for example at certain stages of the tide; or the mangal might provide a key habitat for only certain stages of the life cycle, perhaps as a nursery area for larvae or juveniles.

Fish are abundant in mangrove creeks (Chapter 4). The biomass of fish in creeks and channels within the mangal may be several times as great as that of similar channels in a mudflat area. At high tide, many fish invade the flooded mangrove forest itself, and analysis of stomach contents shows clearly that they are actively foraging for food there (Table 6.3) As many of these commuting fish range over adjacent habitats and inshore waters, they represent an important functional link between mangrove and other habitats.

Among the most important habitats linked with mangroves are seagrass beds (see last section). Given the similarities in the two ecosystems, it is not surprising

Table 6.3 Composition of the gut contents, by volume, of six species of fish caught within a Malaysian mangrove forest to show dependence on plant material and on mangrove animals, data from Sasekumar et al. (1984)

Species	Mangrove	Bivalve	Gastropod	Prawns	Crabs	Fish	Other
Tachyurus	1	3		3	71		21
Arius	1		2	27	46	17	7
Pomadasys			2	16	81		1
Ambassis	44		47		5		46
Liza	38			23			39

that they share some of their fauna, and that faunal movements between mangroves and seagrasses represent a significant functional link, supplementing the coupling of the two ecosystems by movements of POM, discussed above. In particular, many species of fish seem to move freely between seagrass beds and mangrove forests, presumably feeding in both. In the Gazi Bay area of Kenya, $\delta^{13}C$ estimates indicated that the fish species sampled fell into three general categories. One group occurred entirely within the seagrasses, and had tissue $\delta^{13}C$ values that fell entirely within the range for seagrass carbon (-10.07 to -19.82). A second group were found in mangrove creeks, but again had tissue $\delta^{13}C$ values indicating seagrass dependence. The third group of species had $\delta^{13}C$ values intermediate between those of mangrove and seagrass, and presumably had some feeding dependence on both habitats (Marguillier *et al.* 1997).

Larval dispersal and return

Species may also alternate between mangrove and other habitats at different stages of their life cycle. The majority of animals which live as adults within the mangal have larval or juvenile stages which live elsewhere, while the young of some offshore fish and crustacean species use mangroves as a nursery area, for food or for refuge.

The planktonic zoeae larvae of mangrove crabs have already been discussed (p. 98). Vast numbers of these are produced, and released into mangrove creeks or channels, where they are carried away to pass their larval life in open waters nearby. After a series of moults, a zoea metamorphoses into a settling form, the megalopa, then into a miniature crab. Similarly, sessile animals such as barnacles or bivalve molluscs also have planktonic larval stages.

A single female fiddler crab (*Uca*) releases tens or even hundreds of thousands of larvae each year, representing quite a significant export of organic matter from the mangal to nearshore waters. Of these larvae, probably only at most a few hundred survive and return as megalopae. The fact that even this number return, however, is quite remarkable. If the currents carry zoeae passively out to sea, how do megalopae later contrive to return?

The answer is that planktonic larvae are not simply passive organic particles, to be swept wherever current and tide may take them. Larvae in practice often remain not far from, and return to, the area from which they originated. To understand how they achieve this, it is necessary to look more closely at the pattern of water flow in a coastal system such as an estuary.

In an estuary there is a net outward flow of fresh water. In so-called salt wedge estuaries, this tends to move as a surface layer over an underlying wedge of denser sea water. At the interface between the fresh and salt water layers, shear causes a degree of mixing. There are two consequences of this: the salinity of the surface water increases as it moves seawards, and to compensate for the entrained salt water, a slow upstream current of deeper, salt water is generated

Fig. 6.3 Simplified diagram of flow pattern of water within a salt wedge estuary. A landward salt water intrusion is generated beneath the seaward flow of fresh water. (Reprinted from Dame, R.F. and Allen, D.M. (1996). Between estuaries and the sea. *Journal of Experimental Marine Biology and Ecology*, **200**, 169–85, © 1996, with permission from Elsevier Science.)

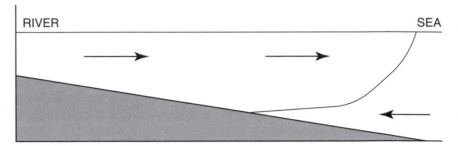

(Fig. 6.3). Where there is a strong tidal flow, more extensive mixing of the two layers takes place, more salt water is entrained, and the upstream current is greater (Dame and Allen 1996).

This combination of downstream and upstream water movements can result in the retention of larvae and their return to a site not far away from their point of origin. Early larvae maintain a position near the surface of the water, and are transported to the mouth of an estuary, or even out to waters of the continental shelf. After a time, they sink, and at deeper levels return upstream by riding the incoming current. Although this transport mechanism has not been demonstrated for any mangrove species, it has been shown to occur with a number of salt marsh species, including fiddler crabs (Epifanio 1988).

Mangroves and fisheries

Shrimps

The relationship between mangroves and other habitats becomes of more than academic interest when commercial fisheries are considered. If species of commercial interest are caught in mangrove creeks or adjacent waters, to what extent do they actually depend, directly or indirectly, on mangrove primary production? Even species caught in offshore waters may rely on outwelled material, or may depend on mangroves for breeding, or as a nursery habitat at certain stages of their life cycles.

This applies equally to fish and to shrimps, although the greatest effort in attempting to establish links with mangroves has probably been with the latter. The shrimp-mangrove relationship also crops up in the context of aquaculture, where in parts of the world large areas of mangrove have been destroyed to construct aquaculture ponds for mass rearing of shrimps. If shrimps depend on mangroves, potential drawbacks to this procedure are perhaps foreseeable. Some of the issues raised by aquaculture are discussed further in Chapter 8.

To clarify one point of terminology: the terms 'shrimp' and 'prawn' are used interchangeably, depending on taste. On the whole, aquaculturists are slightly more likely to use the former, and fisheries scientists the latter, but both terms refer to the same animals, often to the same species.

Most of the commercially fished or cultivated shrimps (or prawns) belong to the family Penaeidae. The geographical distribution of penaeid species fits the geographical distribution of mangroves 'like a hand in a glove' (Chong 1995). The northerly and southerly limits of nearly all penaeids is marked more or less by the 15°C winter isotherm, as is the distribution of mangroves by the 20°C winter isotherm (see Chapter 1). Where there are mangroves there are penaeids. The reverse is not necessarily true: to the north and south of the mangrove zone, penaeids may occur in significant numbers in the absence of mangroves, examples being areas such as Florida, California, the Mediterranean and south-west Australia. Here, however, the place of mangroves may be taken by salt marshes or extensive seagrass beds, so penaeid populations may still be dependent on vascular plants as a primary productivity base.

Areas with particularly well-developed mangroves seem to be associated with particularly rich stocks of shrimp. The islands and channels of the Klang Strait, Malaysia have luxuriant mangrove forests, and the adjacent coastal waters provide an annual harvest of 10 500 t of shrimp (Chong et al. 1996). This, and similar observations, have given rise to several attempts to quantify the relationship between mangroves and shrimps by correlating catches with the area of local mangroves.

In some cases, statistically significant linear relationships have been established. In western Malaysia, for instance, the annual shrimp catch, in thousands of tonnes, was related to mangrove area, in thousands of hectares, by the regression formula

$$C = 0.6368 + 0.5682 \; A$$

In Indonesia, a similar linear relationship was obtained between catch and mangrove area, and in Australia shrimp catch was related to the length of mangrove-fringed coastline, as a substitute for area. More generally, shrimp catch has been related to the area of intertidal vegetation, irrespective of whether this comprised mangroves or salt marsh plants (Turner 1977).

The shrimp catch per unit area tends to decrease at higher latitudes, and this observation has been incorporated into a general formula relating the maximum sustainable yield (MSY) of a shrimp fishery to both mangrove area (A) and latitude (L°):

$$\log_{10} (MSY) = 2.41 + 0.4875 \log_{10} A - 0.0212 \; L$$

Maximum Sustainable Yield is a much-used concept in fisheries science, and refers to the estimated maximum catch that can be harvested without jeopardising the population being exploited (Chong 1995).

What do such correlations mean? The inference that has generally been drawn is that there is a causal connection; that shrimps are so dependent on mangroves that the size of a shrimp population (or annual catch statistics as a substitute for a population estimate) is determined by the area of contributing mangroves. Correlation does not necessarily mean causation, however—as demonstrated by the well-known case of the excellent correlation between the stork population of western Germany and the human birth rate there. Do shrimps actually depend on mangroves, or is the close mathematical relationship a consequence of both being determined by a third variable? Extensive mangroves and sizeable shrimp populations also correlate with large river estuaries, so both could independently be affected by the availability of fresh water, or of silt or nutrient flowing into the sea. What are the connections between shrimps and mangroves?

If there is a direct connection between mangrove area and offshore shrimp catches, it could result from shrimps being dependent on outwelled organic matter, or on the shrimps using the mangal at a certain stage of their life cycle, for example as a nursery area. As discussed earlier, organic matter outwelled from mangroves does not seem to travel very far, and shrimp are caught so far offshore that it seems unlikely that they could directly benefit from it.

Stable isotope studies confirm this. Figure 6.4 shows the distribution of $\delta^{13}C$ values in two species of penaeid, *Penaeus merguiensis* and *Metapenaeus mutatus*, in

Fig. 6.4 Distribution of $\delta^{13}C$ values in four species of penaeid shrimp caught in coastal inlets in a Malaysian mangrove forest (filled bars), compared with individuals of the same species caught 2–18 km offshore (open bars). The range of $\delta^{13}C$ values in mangrove leaf material was −28.5 to −24.5‰. (From data in Rodelli *et al.* (1984). Stable isotope ratio as a tracer of mangrove carbon in Malaysian ecosystems. *Oecologia*, **61**, 326–33, © Springer-Verlag, reproduced with permission.)

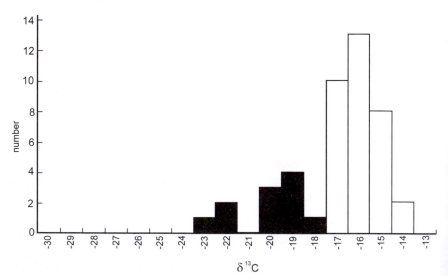

Malaysia. These species were caught both offshore, and in creeks and inlets within the mangal. Bearing in mind the range of $\delta^{13}C$ values of mangrove leaves (see Fig. 4.4), it appears that shrimp within the mangal depend partly, but not entirely, on mangrove carbon. Those collected offshore, in contrast, show no sign of having assimilated carbon directly from mangrove material (Rodelli *et al.* 1984).

The alternative is that offshore shrimps depend on mangroves at an earlier stage of their lives and later move out to offshore waters. As they feed and grow, the mangrove $\delta^{13}C$ signature becomes swamped by later assimilation of carbon from other primary producer sources.

In many species of crustacean, the female carries eggs attached to her abdomen until they hatch as quite advanced larvae, or as juveniles. Penaeids are unusual in this respect. Females may produce as many as 500 000 eggs, and release them directly into the water. Hatching takes place within a few hours, producing larvae called nauplii. The nauplius larva moults several times, before metamorphosing into a different form of larva known as a protozoea. Protozoeae are filter-feeders, dependent on phytoplankton. Neither nauplii nor protozoeae look remotely like adult shrimps. After three or so protozoeal stages, larvae metamorphose again, this time into a more shrimp-like larva, the mysis. After a further three moults (typically, making 11 stages in all) the mysis changes again into a 'postlarva', very similar in body form to the adult shrimp.

Although all penaeids show similar patterns of development, some undergo this elaborate sequence of larval development within an estuary, while in others the life cycle is entirely pelagic. Most species, however, move between offshore and inshore waters at different ages. In many of the commercially important species, spawning takes place at depths of 10 to 80 m, in offshore waters. The larval stages, as always, are planktonic, but postlarvae and juveniles are abundant in mangroves. In some cases, they may actively migrate shorewards. In others, it seems they are carried more or less passively and end up in mangroves as the consequence of current and tidal movements. In the Klang Straits of Malaysia, for instance, it has been demonstrated that strong tidal currents favour the shoreward movement of penaeid larvae, which become trapped in the mangal by physical movement of water along mangrove-fringed channels. Around 65 billion larvae are retained in the Klang mangal by this mechanism (Chong *et al.* 1996).

Virtually all shrimps found in mangrove areas are juveniles, so mangroves clearly merit the term 'nursery areas'. What do young shrimp gain from their residence in the mangal?

The first, and obvious, benefit is nutritional. While in the mangal, juvenile shrimps both ingest and assimilate mangrove detritus: this is confirmed by analysis of gut contents, and by stable isotope studies (Fig. 6.4). The gut contents of Malaysian penaeids may comprise 40 per cent mangrove material, and stable isotope analysis suggests that, on average, perhaps 65 per cent of their carbon originated from mangrove detritus (Rodelli *et al.* 1984). Apart from direct intake of mangrove detritus, mangrove creeks provide a rich habitat within which are many other suitable forms of food for omnivorous shrimps.

Mangroves also provide shelter and protection. At high tide, juvenile shrimps enter the flooded mangrove forest and feed there. When the tide ebbs, they resist the flow until the water level (and available forest habitat) are quite low, then enter the creeks and congregate in turbid shallow waters close to the water's edge at low tide. When the tide rises, the shrimps again move in among the inundated trees. Although this behaviour is related to foraging in the flooded forest, it also reduces the risks of predation. Many fish feed on juvenile shrimp (Table 6.3), and it has been shown experimentally that the structural complexity of dense pneumatophores greatly reduces the efficiency rate of predatory fish, and the mortality rate of shrimps (Primavera 1997*a*).

Fish

Even a few hours with a cast-net fisherman in the mangrove creeks of Malaysia leaves a lasting impression of the richness and diversity of the fish population. Typical, tropical mangroves in the Indo-Pacific region may have close to 200 species. Temperate mangroves, and those in the Atlantic, tend to be rather less species rich, but even here, more than 100 species have been recorded from a single location.

Much of what has been said about the relationship between shrimps and mangroves applies equally to that between fish and mangroves. Analysis of stomach contents confirms the dependence of fish on mangrove invertebrates, and on smaller fish, for food. The majority of mangrove fish are juveniles of species which spawn elsewhere, so that fish clearly use mangroves as nursery areas in much the same way as shrimps do. Finally, the structural complexity of the mangal affords shelter from predators. The abundance and diversity of fish is closely related to the amount of mangrove debris present (Robertson and Blaber 1992). In Sri Lanka, this is exploited by local fishermen, who construct artificial thickets of mangrove branches within mangrove lagoons. After a few weeks these attract and concentrate large numbers of fish, which can then be conveniently trapped.

Fish, like shrimps, depend on mangrove habitats for food, as nursery habitats and for shelter. Many of the species are intrinsically of low commercial value, or are worth little because of their small size. Nevertheless, mangrove fisheries are often of great local significance, and, as with prawns, offshore catches may be of considerable commercial importance (see Chapter 8).

Mangroves and salt marshes

Mangrove habitats are virtually limited to the tropics. In temperate regions, their place is taken by salt marsh vegetation. Apart from their climatic preferences, how do the two ecosystems compare?

Both mangroves and salt marsh plants are adapted to live in frequently inundated, waterlogged, saline and anoxic soils, and both trap and accumulate sediment particles. The most obvious difference between the two is that mangroves are mostly trees, often of considerable height. Salt marsh vegetation comprises grasses and small shrubs, often only a few centimetres in height (although in North America the salt marsh grass *Spartina* may reach 3 m). There is no obvious reason why tolerance of lower temperatures should impose a height restriction. Salt marshes also tend to be less diverse than mangroves: this might be related to the tendency for species richness to decline with increasing latitude, or to salt marshes being less structurally complex than mangroves.

Because the Atlantic coast of North America has a large concentration of both salt marsh habitat (600 000 ha) and biologists, much of our understanding of salt marsh ecology comes from this part of the world, particularly from the dominant salt marsh species of North America, the grass *Spartina*. Primary productivity is comparable with that of mangroves, and may be as high as $500 \text{ g C m}^{-2} \text{ year}^{-1}$. As with mangroves, there is a tendency for productivity to decline with increasing latitude. Productivity also declines with increasing salinity, as it tends to with mangroves. There is some doubt as to whether salt marsh growth and productivity are normally limited by salinity levels, or by limitation of nutrients, particularly of nitrogen (Mann 1982).

The fate of salt marsh production diverges somewhat from that of mangroves. Direct herbivory generally accounts for a greater proportion, more is retained within the ecosystem and less exported (Table 6.4). Leaf shredders do exist, but are less important than, for instance, the sesarmine crabs of the Indo-Pacific mangal. Deposit-feeding fiddler crabs (*Uca* spp.), on the other hand, may be relatively more important. Certainly their abundance is higher: concentrations of up to a staggering 200 crabs per square metre

Table 6.4 Comparison of the fate of primary production in mangroves and salt marshes, from Duarte and Cebrián (1996)

	Mangrove	Salt marsh
Eaten by herbivores	9.1	31.3
Exported	29.5	18.6
Decomposed within system	40.1	51.2
Stored in sediments	10.4	16.7

The figures are in percentages but because each is an average of a number of independent estimates they do not necessarily total 100%.

have been recorded (Macintosh 1982). The burrowing activity of salt marsh *Uca* has a major effect on salt marsh primary production. Experimental reduction of *Uca* population density in one case reduced aboveground production by 47 per cent and increased root mat density by 35 per cent (Bertness 1985).

Finally, salt marshes may be economically as well as ecologically important habitats, and for similar reasons to mangroves. They play a major role in protecting against coastal erosion, and sustain coastal shrimp and other fisheries (Turner 1977).

7 Biodiversity and biogeography

This chapter considers the intersecting questions of biodiversity and biogeography. What determines the diversity and distribution of species of mangrove flora and fauna throughout the world?

Biodiversity is an elusive concept. Before considering how and why it varies, and the significance of that variation, the concept must be explored and, if possible, clarified. Important questions of scale are involved: biodiversity can be considered at many scales, from microscopic to global.

As discussed in Chapter 1, mangroves are restricted by temperature, being almost exclusively tropical and subtropical in geographical distribution. This overall distribution comprises the separate ranges of many individual species. These cannot be explained purely by physical or climatic limitations, if only for the obvious reason that similar climatic regions in different oceans are occupied by different species of mangrove. The biogeography of mangroves raises questions of the origin and spreading of species: when and where did mangroves originate, and what factors led to their present distribution? Much of the present distribution pattern of mangrove species can be explained only by taking account of evolutionary events: the past is the key to the present.

The actual number of species of mangrove tree (and of mangrove-associated plants and animals) differs, not just geographically, but between sites within a geographical region. To what extent does biodiversity depend on biogeographical or evolutionary factors, and to what extent on local physical or biotic conditions?

And does biodiversity actually matter? Theoretical considerations suggest that biodiversity might have important influences on ecosystem function and that, in particular, diversity and productivity are linked. What is the functional significance of mangrove biodiversity?

What, if anything, is biodiversity?

'Biodiversity' has become common currency with the media and with politicians. In the process, a complex concept has come to be used (and misused) in an oversimplified way. It has been caustically defined as 'biological diversity—with the logical part taken out'. What *is* biodiversity?

In its simplest and most straightforward sense, it is often taken to mean simply the number of species present. Even this sense needs qualification. The area in question must be defined: all other things being equal, the greater the area being considered, the greater the number of species that will be encountered within it. Species–area relationships have been shown for many groups of organisms, over a wide range of spatial scales. The precise relationship between species and area— typically a simple exponential one—is much more informative than simple species catalogues, but in most cases the information has not been assembled in an appropriate way, and species listings must suffice (Rosenzweig 1995).

Even simple species counts have hidden complications. Relative abundance of species in some sense contributes to diversity. Two patches of habitat might each contain 100 individuals. In one, the 100 individuals comprise 10 of each of 10 different species. In the other, 10 species are also present: in this case one species dominates, accounting for 91 of the 100 individuals, with the other species each represented by a single individual. The second habitat is surely less diverse than the first. A further complication arises if taxonomic levels are taken into account. A habitat containing 100 species of insect, all of which are beetles, is less diverse than one with 100 species of widely different affinities. Finally, of course, intraspecific diversity, particularly genetic variation within a species, is of enormous evolutionary importance.

Species diversity at a specific point is sometimes referred to as alpha diversity. Introducing a spatial dimension and recording the accumulation of species with distance as we move away from a point has been termed beta diversity. Species– area curves are one method of measuring beta diversity. Gamma diversity refers to the number of species in a whole region. These terms were in vogue for a time, and are still encountered occasionally, but seem now to be regarded as unnecessary jargon (Rosenzweig 1995).

As far as mangrove biodiversity is concerned, pragmatism dictates the basic estimation of species number as the most useful approach.

Regional diversity

The geographical distribution of mangrove species, and of mangrove habitats, is limited by temperature. More specifically, the 20°C winter sea temperature isotherm, with few exceptions, circumscribes the range of mangroves through- out the world (Fig. 1.1). Possible reasons for this limitation have been discussed earlier (p. 32).

If the latitudinal range of mangroves is curtailed by temperature, the long- itudinal distribution of species is affected primarily by physical barriers. The tropical coastlines of the world are separated by the great land masses, and by open oceans, into two major regions and a number of smaller ones. From the point of view of mangroves, the primary separation is between the Indo–West Pacific (IWP) and the Atlantic–Caribbean–East Pacific (ACEP) regions. Africa, and the cold waters round the southern tip of Africa, effectively prevent contact between western Indian Ocean and eastern Atlantic tropical waters.

The barrier between the western and eastern Pacific is less obvious. It consists, quite simply, of the largely empty space of the central Pacific, which forms a barrier to mangrove dispersal. The number of species of mangrove declines eastwards across the Pacific, the easternmost limit being Samoa (170°W). Beyond this, no mangroves occur naturally until the west coast of the Americas. If artificially introduced to intermediate islands (for example to the Society Islands or Hawaii) they thrive, showing that their absence is due to inaccessibility, not to a lack of suitable habitats. A single species, *Rhizophora samoensis*, appears to have crossed the Pacific in the reverse direction, spreading from the Pacific coast of South America as far as New Caledonia (Fig. 7.1). It is virtually indistinguishable from the American species *R. mangle*, presumably a recent ancestor (Tomlinson 1986; Woodroffe 1987).

The ACEP region falls into three subregions: the eastern Pacific; the Caribbean and western Atlantic, separated from it by the American continent and the Isthmus of Panama; and the Atlantic coast of West Africa, separated from the western Atlantic by open ocean. Similarly, the IWP region is biogeographically separated into East Africa (and the Red Sea); an 'Indo–Malesian' subregion comprising the Indian subcontinent and west and South-east Asia; and

Fig. 7.1 The distribution of mangrove species (and recognised hybrids) in the western Pacific. (Reproduced with permission from Woodroffe, C.D. (1987). Pacific island mangroves: distribution and environmental settings. *Pacific Science*, **41**,166–85, © University of Hawaii Press.)

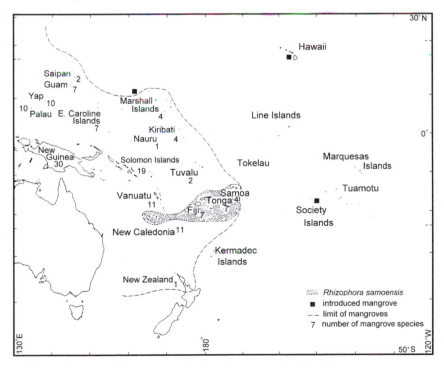

Australasia, consisting of Australia, New Zealand, New Guinea and the islands of the western Pacific. These six subregions are shown in Fig. 1.1.

The IWP and ACEP regions have comparable total areas of mangrove habitat, but widely different numbers of genera and species (Table 7.1). Clearly mangrove species are not evenly distributed around the tropics. Many more species and genera are present in the IWP, particularly in the Australasian and Indo–Malesian subregions, than in the ACEP region.

The contrast between the two major regions is highlighted in Table 7.2. It is clear that the Indo–West Pacific contains not just different species and even genera of mangroves, but many more of them: more than three times as many genera (23 compared with 7), and nearly five times as many species (58 compared with 12). Depending on whether certain species are classified as true mangroves, or mangrove associates (Chapter 1), these figures can be made to differ slightly, but the general conclusion is always that the IWP contains substantially more species and genera of mangrove than the ACEP (Ricklefs and Latham 1993).

Figure 7.2 also compares the relative similarity of the subregions with respect to the actual species present. The six subregions fall into two distinct clusters, confirming the overall division into IWP and ACEP regions. No species are found in both ACEP and IWP regions, apart from the mangrove fern *Acrostichum aureum*. *Avicennia marina* (an IWP species) has recently been artificially introduced to California, and the mangrove palm *Nypa fruticans* to West Africa and to the Atlantic coast of Panama.

Table 7.1 Area of mangrove habitat, and genera and species of mangrove, found in the different zones of the Indo-West Pacific (IWP) and Atlantic–Caribbean–East Pacific (ACEP) regions, from Ricklefs and Latham (1993)

Region and subregion	Mangrove area (km^2)	Genera	Species
IWP			
Australasia	17 000	16	35
Indo–Malesia	52 000	17	39
East Africa	5000	8	9
ACEP			
West Africa	27 000	3	5
West Atlantic–Caribbean	48 000	3	6
Eastern Pacific	19 000	4	7

Table 7.2 Number of genera, and of species, exclusive to the Atlantic–Caribbean–East Pacific (ACEP) and to the Indo–West Pacific (IWP) regions, and the number common to both regions, data from Duke (1992)

	ACEP only	Both	IWP only
Genera	4	3	20
Species	11	1	57

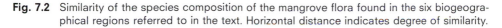

Fig. 7.2 Similarity of the species composition of the mangrove flora found in the six biogeographical regions referred to in the text. Horizontal distance indicates degree of similarity.

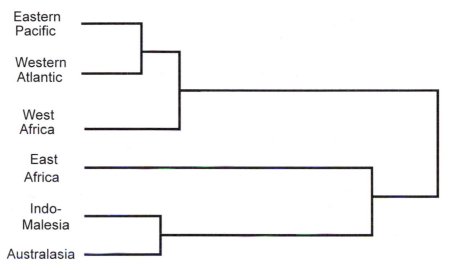

Only the two genera containing most species, *Avicennia* and *Rhizophora*, and the mangrove fern *Acrostichum*, occur naturally in both regions. Two sister genera, *Laguncularia* and *Lumnitzera*, probably separated relatively recently from a cosmopolitan ancestor: the former genus occurs only in the ACEP region, the latter in the IWP (Duke 1992; Ricklefs and Latham 1993).

The number of species of mangrove-associated animals probably also tends to be significantly greater in the Indo–West Pacific, although this is much affected by collector bias. For some reason not every scientist is equally fascinated by the taxonomy of, say, polychaete annelids. As a result, the number of species recorded at a site depends heavily on whether a polychaete expert happens to have visited or not.

Among the most important groups of mangrove animal, there are certainly more species of sesarmine crab in the IWP: at least 37 species recorded in Australia and 44 from Malaysia (and perhaps a further 20 per cent still to be described), compared with eight or so from West Africa. Some 211 species of mollusc have been recorded from the Indo–Malesian area, and 145 from Australasia, compared with 124 from the West Atlantic/Caribbean. With due scepticism about some of the evidence, it can probably be concluded that IWP mangrove habitats are richer in at least crabs and molluscs, both key groups in mangrove ecology, although the effect is less marked than with mangrove trees themselves (Ricklefs and Latham 1993).

The origin of the disparity in species between the two major biogeographical regions is discussed in the next section.

Origins

How has the present distribution of mangrove species come about? One clue comes from the observation that more genera and species of mangrove occur in South-east Asia than elsewhere in the world. Figure 7.3, for instance, shows the geographical distribution of the four related mangrove genera of the family Rhizophoraceae, and similar distribution maps could be drawn for a number of other mangrove genera. A traditional explanation has been that the South-east Asian centre of diversity represents also a centre of speciation: that mangroves originated here and disseminated eastwards across the Pacific, to the western coast of the Americas, and westward to East Africa, and then to the east and west coasts of the Atlantic. The expansion of a species would be limited by the dispersal properties of its propagules, by the distribution of suitable habitats, and by major physical barriers. As not all species were equally good at dispersing, diversity would tend to decline with increasing distance from the source.

It is hard to reconcile an eastward spread across the Pacific to the Americas with the absence of mangroves from many of the Pacific islands (Fig. 7.1), and equally hard to see how mangroves could have spread round the southern tips of either Africa or South America. In fact the current distribution of mangroves cannot be understood in relation only to the present configuration of the land masses of the world. The present can be explained only by the past: by the chronology of mangrove evolution, and by the movement of continents.

Mangroves have a long and conservative fossil history. There are many problems in interpreting fossil evidence. Apart from the possibility of misidentification of poorly preserved remains, fossil material can often be assigned only to a genus, or even family. Not all present-day members of the family Rhizophor-

Fig. 7.3 Distribution of four mangrove genera, *Kandelia, Rhizophora, Ceriops* and *Bruguiera*, of the family Rhizophoraceae. (Reproduced with permission from Ricklefs, R.E. and Latham, R.E. (1993). Global patterns of diversity in mangrove floras. In *Species diversity in ecological communities* (ed. R.E. Ricklefs and D. Schluter), pp. 215–29, © 1993 by the University of Chicago.)

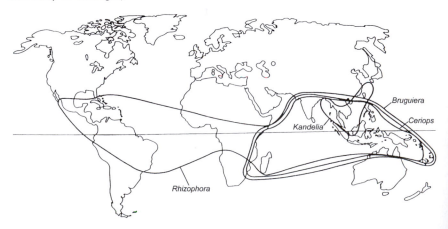

aceae, for example, are mangroves, so fossil evidence of the existence of member of this family is not evidence for the existence of mangroves. Even identification in the fossil record of a genus all of whose modern members are mangroves is not conclusive. Ideally, corroborative evidence of a mangrove environment is desirable, such as fossil molluscs identifiable with modern mangrove residents. Conversely, of course, a fossil may not be recognised as mangrove if it belongs to taxonomic group that is now extinct, or which has no modern mangrove representatives.

Fossils interpreted as mangroves have been reported from the early Cretaceous (Fig. 7.4), or even earlier, but most are at best doubtful. The earliest reliable

Fig. 7.4 The record of fossil pollen of five families containing mangrove species. The families include non-mangrove members: the horizontal bars indicate the oldest records of mangrove genera within each family. Figures are the approximate start of the geological period in question, in millions of years before the present. (From Duke, N.C. (1995). Genetic diversity, distributional barriers and rafting continents—more thoughts on the evolution of mangroves. *Hydrobiologia*, **295**, 167–81. Reproduced with kind permission from Kluwer Academic Publishers.)

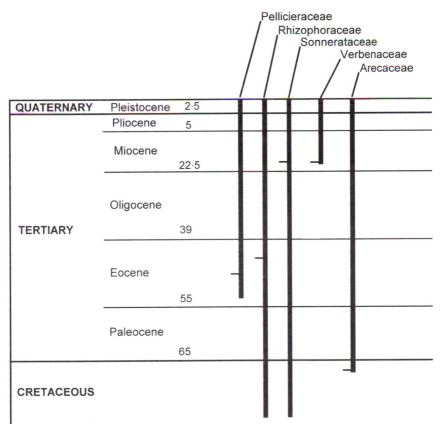

mangrove fossil known is of fruit of the palm *Nypa*, from the mid-Cretaceous period. Pollen fossilises well, and is distinctive enough for identification to genus, sometimes even to species. The oldest mangrove pollen known is that of *Nypa*, from the end of the Cretaceous period, around 69 million years ago, and the early Palaeocene. Palaeocene *Nypa* pollen has been found in North and West Africa and Eastern Brazil, as well as in South-east Asia. Thereafter, *Nypa* is known from Eocene deposits in the Caribbean, South America, Africa and Europe, as well as Asia. By the Miocene it appears to have retracted to Asia, and currently, apart from deliberate reintroduction in the Caribbean and West Africa, *N. fruticans* is limited to Sri Lanka, South-east Asia, the Philippines, Australia and New Guinea (Fig. 7.5) (Duke 1995; Plaziat 1995).

Other mangrove species show a broadly similar pattern. Fossils assigned to *Rhizophora, Ceriops, Bruguiera, Avicennia* and *Pelliciera* have been discovered in English, French and Spanish strata of the late Palaeocene, Eocene and Miocene, in some cases with corroborative evidence such as traces of oysters having attached to prop roots (p. 78). In addition, two extinct genera, *Wetherellia* and *Palaeowetherellia*, were probably mangroves; they have been identified in marine Eocene deposits in Maryland, Germany and England, in association with *Nypa* (Plaziat 1995; Ricklefs and Latham 1993).

This distribution makes no sense in relation to modern geography. At the time, however, Africa was separated from the land mass of Europe and Asia by a major sea, known as the Tethys, which connected the Indian Ocean, through what is now the Mediterranean, to the forerunner of the modern Atlantic. North and South America were not joined at the Isthmus of Panama, so that the Pacific, the Tethys, and the Indian Oceans formed a continuous sequence.

Fig. 7.5 Distribution of fossil evidence of the geographical distribution of the mangrove palm *Nypa*, compared with its current distribution (recent deliberate introductions excluded). (Reproduced with permission from Ricklefs, R.E. and Latham, R.E. (1993). Global patterns of diversity in mangrove floras. In *Species diversity in ecological communities* (ed. R.E. Ricklefs and D. Schluter), pp. 215–29, © 1993 by the University of Chicago.)

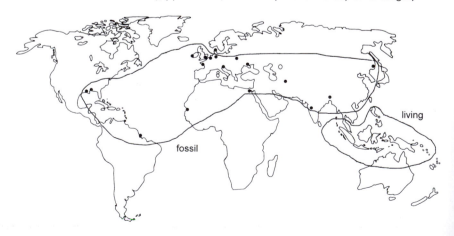

The present-day land barriers of North and South America, and Africa with Asia, did not exist, while America and Africa had not drifted far enough apart for the young Atlantic to limit dispersal. There was no impediment to the spread of mangroves along the margins of warm seashores, virtually around the world (Fig. 7.6).

Thereafter, the latitudinal ranges of many species contracted, probably due to a general cooling of the earth. In Europe and North Africa, mangroves disappeared completely; in Central America, some species dwindled while others vanished. Then, around 30 to 35 000 000 years ago, closure of the Tethys separated the Indo–West Pacific region from the Atlantic, while widening of the south Atlantic completed a barrier between the mangroves of the eastern and western Atlantic margins. North and South America united for the last time, the Isthmus of Panama finally separating the eastern Pacific from the western Atlantic species. This took place a mere 2 to 3 000 000 years ago, so that there are few differences in the mangrove floras on either side of the Isthmus (Duke 1995).

Tectonic movements are therefore responsible for the segmentation of the originally continuous mangrove distribution. Following the separation, regional extinctions and diversification of species proceeded. As the outcome is many more genera and species in the Indo–West Pacific than in the Atlantic–Caribbean–East Pacific regions, one must conclude that fewer species became extinct, or that more evolved, in the IWP. It is virtually impossible to resolve the issue, but the balance of opinion is that the wet climates more widespread in the IWP offered more scope for terrestrial species to adapt to the mangrove habitat along a gradual salinity gradient. Arid climates—such as appear to have been typical of Africa and the New World tropics at the periods of major

Fig. 7.6 Position of the major land masses during the Eocene period. The symbols indicate the locations of mangrove fossils from the period: simple star, *Nypa*; star in circle, other mangrove species. (Reproduced with permission from Plaziat, J.C. (1995). Modern and fossil mangroves and mangals: their climatic and biogeographic variability. *Geological Society Special Publication*, **83**, 73–96, © Geological Society.)

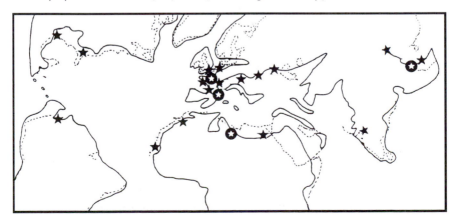

speciation of mangroves—are intrinsically less favourable for mangroves, and tend to produce an abrupt transition with a zone of high salinity at the upper intertidal level of the shore. This might have inhibited the emergence of new mangrove species by extension of terrestrial species into the intertidal (Ricklefs and Latham 1993).

Local diversity

The species present at a specific location are a subset of the species present in the region as a whole. However, the local diversity of species—at a scale of, say, hectares—need not follow regional patterns. It would be perfectly possible for a region to contain many species, each with a narrow geographical distribution within the region. The result would be only few species present at any one location. In fact the pattern of local diversity of mangrove species does follow regional trends. Not only does the ACEP region have fewer mangrove species overall, but locations within the region typically contain only three to four species. This represents about half of the total regional pool of species. As not every ACEP species is geographically widespread within the region, this means that virtually all of the available species are represented at each location. This contrasts further with the IWP, which has many more species overall and greater heterogeneity between sites in species present (Ricklefs and Latham 1993).

What determines the number of species at a particular location, as opposed to the species richness of an entire region? In general local processes such as competition tend to reduce species number: regional processes tend to increase local diversity through the movements of individuals and the resulting spread of species between habitats and habitat patches (Ricklefs and Schluter 1993).

Similar species at the same trophic level—such as the species of mangrove tree—compete for similar resources. Interspecific competition may have two outcomes. On the one hand, less successful competitors may simply disappear. Alternatively, competing species may minimise the cost of competition by becoming more specialised. With mangroves, this might manifest itself in the occupation of narrower segments of a salinity gradient, or of tidal level. Conversely, where an ecosystem contains fewer competing species, the outcome might be niche expansion, with species being more broadly distributed in relation to clinal environmental variables.

The ACEP has fewer species than the IWP, locally as well as regionally. If less intense interspecific competition favours broader ecological niches, ACEP species should, in general, show expanded niches: they should be found across a wider range of environmental variables than IWP species.

With this in mind, the distribution of ACEP and IWP mangrove species within sites can be compared. Table 7.3 shows where species are found with respect to two variables, height in relation to tidal range and distance up or down an estuary. Because the data were assembled from many published sources, where

Table 7.3 Occurrence of mangrove species at different estuarine locations and tidal levels for IWP and ACEP species, from data in Duke (1993)

	Downstream	Intermediate	Upstream
Indo–West Pacific (IWP)			
high	11 (21)	20 (38)	9 (17)
mid	15 (23)	23 (43)	9 (17)
low	9 (17)	10 (19)	6 (11)
Atlantic–Caribbean–East Pacific (ACEP)			
high	5 (42)	5 (42)	2 (17)
mid	5 (42)	5 (42)	1 (8)
low	2 (17)	1 (8)	0 (0)

Figures in brackets are percentages of the total species in the region for which data was available. Estuarine locations were divided arbitrarily into upstream, intermediate and downstream thirds of estuary.

different classifications of tidal range were applied, tidal height is divided simply into high, middle and low. This reflects inundation frequency, a key variable in mangrove distribution (Chapter 1). Estuarine position was arbitrarily divided into downstream, intermediate and upstream. This determines the mix of river and sea water to which a tree will be exposed, and therefore reflects salinity exposure, another key environmental variable. The matrix combining the two therefore gives an impression of overall niche breadth, at least as far as these two environmental variables are concerned. In the Table, progression from the upper right corner (upstream in the estuary, high tidal position) to the lower left (downstream, low tidal position) could be regarded as a reflection of a shift towards more advanced mangrove habits, with greater tolerance of frequent inundation and higher salinity.

Mangrove species in the ACEP region do not appear to show broader ecological distributions. Indeed one sector—upstream, low tidal position—is apparently devoid of mangrove species, although in the IWP several species occupy this ecological position. Moreover, ACEP species are no more likely to spread into more than one of these ecological categories. In both regions, 42 per cent of the species occur in only one of the nine possible combinations of estuary and tidal levels, and 58 per cent in more than one (Duke 1993).

There is therefore no clear evidence that processes such as interspecific competition play a major part in determining the number of mangrove species found at a particular location.

Stochastic factors may also affect the number of species of mangrove present. It is well known that the number of species present on an island is often related to the area of the island (Rosenzweig 1995). An island, in this context, can be any area of suitable habitat separated from other such areas by stretches of unsuitable habitat: it need not necessarily be a physical island surrounded by sea. The species–area relationship results largely from a balance between the successful establishment of new colonisers and the chance extinction of established species, both being affected, differently, by the available area. With mangrove species, establishing any such relationship is complicated by many other factors, such as defining 'islands'. In some cases, however, it is possible to see how area might locally affect species number. Mangrove species numbers, plotted against total island area, are shown in Fig. 7.7, for a group of four adjacent West African islands, Bioko (Fernando Po), Principe, Saõ Tomé and Annobon.

To be present on an island, a species must first get there: floating propagules must have arrived in sufficient numbers to establish a founding population. The longer the time at sea, or the greater the distance from source, the lower the chances of success. Figure 7.8 shows the relationship between mangrove species number and distance from the mainland of Africa to each of the same four islands. For comparison, the nearest country of the African mainland, Cameroon, has been added to the graph. Species number is clearly related, inversely, to distance from the mainland. (A similar relationship appears if distance from the nearest landward island is plotted instead of distance from the mainland, on the assumption that mangroves might have used each successive island as a stepping stone to the next.) This relationship is much the same as that between mangrove species on various islands of the Pacific and distance from Australia or New Guinea (Fig. 7.1), but at a distinctly smaller scale.

The species most widely dispersed across these islands should be those whose propagules last longest at sea. In general, this appears to be the case. *Avicennia germinans* and *Rhizophora harrisonii*, with propagule survival times estimated at

Fig. 7.7 Mangrove species number in relation to area (km^2) for the four West African islands of Bioko (Fernando Po), Principe, Saõ Tomé and Annobon. (Reproduced with permission from Saenger, P. and Bellan, M.F. (1995). The mangrove vegetation of the Atlantic coast of Africa. A review. pp. 1–96. Toulouse, Laboratoire d'Ecologie Terrestre.)

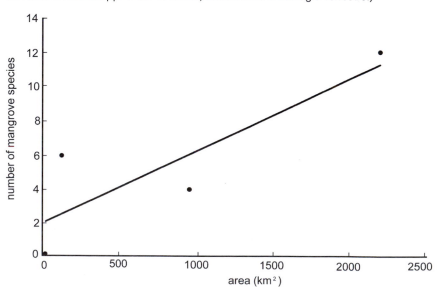

Fig. 7.8 Mangrove species number in relation to distance from the mainland for the four West African islands of Bioko (Fernando Po), Principe, Saõ Tomé and Annobon, and for Cameroon. (From data in Saenger, P. and Bellan, M.F. (1995). *The mangrove vegetation of the Atlantic coast of Africa. A review.* pp. 1–96. Toulouse, Laboratoire d'Ecologie Terrestre. With permission.)

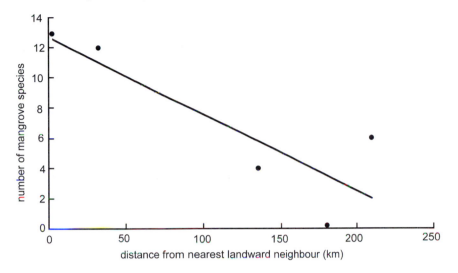

3 months and more than a year, respectively, are found on all three of the mangrove-occupied islands (as well as on the adjacent mainland). *Laguncularia racemosa*, which survives only a month at sea (Table 1.6), is restricted to the island nearest to the mainland, Bioko (Fernando Po). Surprisingly, however, the two other species of *Rhizophora*, *R. mangle* and *R. racemosa*, are restricted to the African mainland and the nearest island, Bioko. As far as is known the propagules of these species have similar dispersal ability to those of *R. harrisonii*.

Genetic diversity

Biodiversity can also be considered at a level below that of the species. The extent of genetic differences between individuals within a population is an important measure of the diversity of that population. Assessment of the genetic differences between populations of the same species gives a measure of the extent to which such populations are genetically isolated from each other, while genetic comparison of different species can give information about the extent of divergence and, potentially, when such divergence took place.

Among the red mangrove of Florida and the Bahamas (*Rhizophora mangle*), studies of chlorophyll-deficient mutations indicate that the population relies predominantly on self-pollination: more than 95 per cent of embryos appear to result from selfing. As a result of inbreeding, the populations contain relatively little genetic variation.

Rhizophora tends to form dense, monospecific forests. Agricultural ecosystems are also genetically uniform monocultures and, as a result, are often particularly prone to disease and pest infestations. This suggested that the low species and genetic diversity of mangroves might be a factor in the occasional mass defoliation episodes which have been recorded (p. 55).

However, such high rates of self-pollination, and low genetic diversity, are not universal. The same species of mangrove in Puerto Rico had a lower rate of self-pollination (71.2 per cent), so the mode of pollination adapts to local circumstances rather than being a fixed species-specific characteristic. Populations with more frequent self-pollination had higher mutation rates, and one intriguing possibility is that an increased mutation rate has evolved to offset the low genetic variation resulting from inbreeding (Klekowski *et al.* 1994).

DNA polymorphism can now be assessed directly, so that genetic analysis is no longer restricted by the luck of finding appropriate mutations (and the virtual impossibility, with mangroves, of carrying out significant breeding programmes). Molecular genetics studies on *Avicennia* in India have been used to assess and compare the amount of genetic variation within separate populations, between populations of the same species and between species. *Between* populations, genetic variation is relatively high, suggesting that the separate populations are largely isolated from each other, with little mutual gene flow. The amount of genetic variation *within* a population varies from one population to another, suggesting that it is dictated by local circumstances. Finally, comparisons between species of *Avicennia* (*A. marina*, *A. alba* and *A. officinalis*) indicate that *A. marina* is closer to *A. alba* than to *A. officinalis*, and presumably diverged from it more recently (Parani *et al.* 1997).

These examples give some idea of the potential of genetical analysis in understanding the population biology of mangroves. Genetic studies of mangroves are in their infancy, but have already provided some useful information and promise much more, even if so far they have raised more questions than have been answered.

Diversity of the mangrove fauna

Local species richness of the mangrove fauna is also affected by factors such as habitat area, dispersal ability and distance from areas of similar habitat. By far the most detailed studies are the classical experiments of Simberloff and Wilson on some small mangrove islands among the Florida Keys (Simberloff 1976; Simberloff and Wilson 1969, 1970).

Firstly, they established an inventory of the species of terrestrial invertebrate (principally insects) present on each of a series of small mangrove islets, which ranged from 11 to 25 m in diameter. The islets were then handed over to professional pest-controllers, who erected scaffolding, encased each islet in plastic sheeting, and proceeded to fumigate with methyl bromide, killing all terrestrial arthropods present. This was not, perhaps, the most environment-friendly of experiments: however the fumigated islets were extremely small (some little more than a single tree) and, as will soon be apparent, recovered rapidly.

Periodically, over the ensuing 2 years, the recolonising species present were recorded. In each case, the number of species rose, then levelled off, roughly asymptotically, at a number of species close to the number present before the fumigation (Fig. 7.9).

This is more surprising than it seems. Although the total number of species present was about the same as that before the fumigation, the actual species present were not the same ones as before; nor were they necessarily the same from one census to the next. As an example, seven species of Hymenoptera (ants, bees and wasps) were present on one islet before fumigation, and eight species a year after: but only two of the species were present on both occasions. Species were therefore continually being lost, and gained, and the total present at any one time was a 'snapshot' of a dynamic equilibrium.

What determines the equilibrium species richness for one of these mangrove islets?

Fig. 7.9 Number of species of terrestrial arthropod present following fumigation of four mangrove islets in the Florida Keys. The number of species identified at each islet before fumigation are indicated on the vertical axis. The islets were similar in area, but differed in distance from the nearest possible source of immigrants. The islet with the lowest number of species before fumigation, and after recovery, was furthest from a source, and that with the greatest number of species before and after the experiment was closest to a source. (Reproduced with permission from Simberloff, D.S. and Wilson, E.O. (1970). Experimental zoogeography of islands. A two year record of colonization. *Ecology*, **51**, 934–7, © Ecological Society of America.)

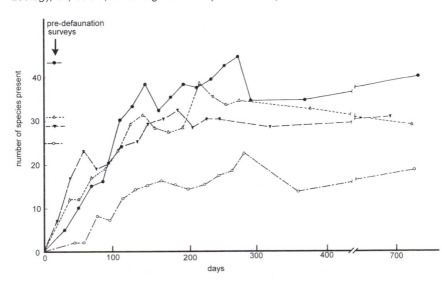

As with tree diversity, discussed above, the number of species of terrestrial invertebrate on Florida Keys mangrove islands correlates inversely with distance from source habitats, and directly with the island area.

The problem with simply correlating area with species number is that larger areas may incorporate greater habitat diversity than smaller ones. A close species–area relationship does not allow the inference that area determines species number, since a high level of diversity might result from greater habitat complexity, rather than from greater area.

Simberloff and Wilson ruled out this possible confusion of variables by artificially reducing the area of a group of mangrove islets with power saws. This had the effect of reducing area without any concomitant change in habitat complexity of the surviving vegetation. The result was that invertebrate species diversity fell, demonstrating its dependence on habitat area (Fig. 7.10).

Terrestrial arthropods and isolated mangrove islets obviously constitute a special case. Most mangroves are much less isolated from sources of colonising species, either terrestrial hinterland or the sea. With larger and longer-lived species than the insects and other terrestrial arthropods, and with larger areas of mangrove than these tiny islets, the dynamics of the species turnover will in any case be vastly different. Nevertheless, Simberloff and Wilson's experiments make the crucial point that the species composition and richness of the mangal is not necessarily static, or even the result of wholly deterministic processes.

Diversity and ecosystem function

It seems reasonable to suppose that the species-richness of an ecosystem will have some bearing on its function. In particular, it has been postulated, largely on theoretical grounds, that a diverse ecosystem will be more productive, or more resilient, than a species-poor one. The precise reasons for these expectations need not concern us, but clearly the predictions could be tested.

The problem is that real ecosystems are generally too complex, and subject to too many environmental influences, to give a clear answer. Ecologists have therefore resorted to experiments with artificial ecosystems, with contrived levels of diversity and mixtures of species. These have included systems as different as grassland plots, small-scale ecosystems set up in artificial microcosm containers and assemblages of micro-organisms (McGrady-Steed *et al.* 1997; Naeem and Li 1997; Tilman *et al.*1997).

Although the results are complex, some general conclusions have emerged. Adding more species tends to increase productivity of the system as a whole—but only up to a certain limit. Further increases in species number have a greater effect on ecosystem stability: the more species present, the smaller the fluctuations with time, and differences between replicates, in measures of ecosystem function as a whole.

What initially increases productivity is probably the inclusion of additional

Fig. 7.10 Individual species/area curves for mangrove islets whose area was artificially reduced. Points relating to each islet are joined. The islet designated IN1 was a control, and the change in species number presumably reflects random change. (From Simberloff, D. (1976). Experimental zoogeography of islands: effects of island size. *Ecology*, **57**, 629–48, © Ecological Society of America.)

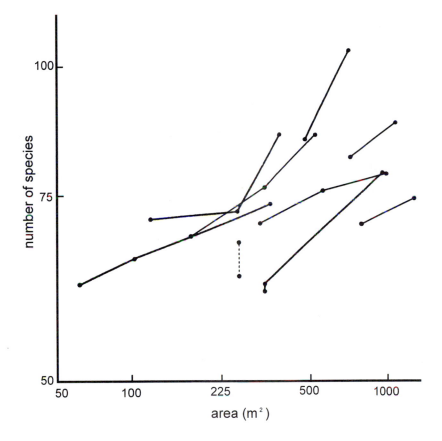

functional groups of organisms. (A functional group is a particular mode of exploiting resources within the environment. Trees, shrubs and epiphytic algae, for instance, all operate at the same trophic level, but in very different ways.) Once an appropriate mix of functional groups has been reached, adding further species may intensify competitive interactions between species, but will not significantly enhance productivity of the ecosystem as a whole. A further increase in species richness may increase stability and predictability of the ecosystem without a concomitant increase in productivity.

Natural ecosystems are generally so rich in species, and affected in such complex ways by environmental variables, that there is little scope for testing the relationship between diversity, productivity and stability. Mangrove ecosystems may offer a system simple enough to address the question of the functional significance of diversity. Although the mangrove fauna is relatively rich in species, the habitat is generally dominated by a small number of tree species, and is structurally simple compared with other tropical, or even temperate, forests. Moreover, mangrove forests occur in similar environmental settings in many different parts of the tropics, but the number of tree species varies greatly between different biogeographical regions. It is therefore possible to find comparable mangrove ecosystems, in similar environmental conditions, which differ principally in tree species richness, and to test whether any differences in productivity can be related to differences in diversity.

If productivity was directly related to species richness, mangrove habitats of the Indo–West Pacific (IWP) should be more productive than those of the Atlantic–Caribbean–East Pacific (ACEP). This appears not to be the case. Productivity and standing crop biomass vary considerably between sites, and tend to decrease with increasing latitude, but a comprehensive collection of the available data shows no particular differences between biogeographical regions (Saenger and Snedaker 1993). Such a general comparison masks local differences in many factors known to affect productivity, such as environmental setting and salinity gradients. A better 'experiment' would be to compare the productivity of mangroves in a large-scale transect from west to east across the islands of the Pacific, where climate and substrate are similar but species number, limited by dispersal ability, varies from 30 in New Guinea to a single species in some of the more outlying islands (Fig. 7.1) (Woodroffe 1987).

Mangroves tend to arrange themselves into relatively homogeneous, almost monospecific patches or bands, often aligned with physical gradients of the environment (Chapter 2). At a scale of hectares, a mangrove forest may encompass many species and be regarded as relatively diverse. At a scale of square metres, in contrast, diversity appears low, and a sample may include only a single species. Tree diversity may not appear to correlate with productivity because at the spatial scale relevant for species interactions the species scarcely intermingle.

In one case, the spatial scales of diversity and productivity measures are compatible. This is the 'experiment' of the Matang, a managed mangrove ecosystem in western Peninsular Malaysia, described more fully in the next chapter (p. 175). One effect of the management regime, over the last 90 years or so, has been to reduce tree diversity, virtually to a single species. The productivity appears to be declining. A clear-felled patch of forest in the 1970s yielded 136 t ha^{-1}, while a decade previously the yield had been 158 t ha^{-1}. However, there is no particular reason to attribute this apparent decline in productivity to a reduction in tree diversity, rather than to other factors such as the effect of invading *Acrostichum* reducing seedling growth (see next chapter), or to minor alterations in management practice (Gong and Ong 1995).

Adult mangrove trees may not always interact much in life, but their principal contribution to ecosystem processes, shed leaves, intermingle on the forest floor. Mixed-species leaf litter decomposes no faster than leaves from a single species, suggesting again that tree diversity has little effect on the rate of energy flow through the ecosystem.

Litter processing could also be affected by the diversity of either leaf-processing crabs (p. 83), or of decomposing micro-organisms. Crab abundance certainly affects litter processing rates, but the impact of crab diversity has not been investigated. Microbial diversity is a problematic concept. Diversity is generally extremely high, with many species globally distributed: 'in the case of micro-organisms, 'everything is everywhere" (Fenchel *et al.* 1997). In natural circumstances, microbial biodiversity probably does not vary sufficiently to affect mangrove ecosystem processes.

With the possible exception of the Matang, there is little evidence that variations in biodiversity—whether of mangrove trees, the mangrove fauna or of micro-organisms—has any significant effects on ecosystem productivity.

Whether or not it affects productivity, tree diversity may also be relevant to the long-term stability of a mangrove forest. In many parts of the world, mangroves are subjected to intermittent catastrophic damage from typhoons or hurricanes. These create gaps, often filled by different species from the ones that have been destroyed. A forest with a mix of species, some more resistant to hurricane damage, and others better at rapidly occupying gaps, may be more stable over a time scale of decades. Here, too, the significant factor may be functional differences between species, rather than actual species diversity. Species-rich IWP mangrove forests should recover more rapidly than relatively species-poor ones in the ACEP region, but it is difficult to arrange for comparable forests to be visited by appropriately matched hurricanes in the two regions.

8 Impacts

By far the most ecologically significant animal associated with mangroves is the human species. Humans benefit in many ways from mangrove ecosystems, deriving products such as timber and fuel wood, and services such as protection against coastal erosion. Some of these are reviewed in this chapter.

Inevitably, significant exploitation of mangroves means that few mangrove forests are now pristine: most have been affected to some degree. Any extraction removes material which would otherwise have contributed to the ecosystem. Selective removal of certain size categories of tree for timber affects the age structure and architecture of the forest, and selective removal of some species in preference to others affects the community structure and reduces diversity. Physical disturbance during the extraction process may prevent the establishment of seedlings to replace the trees removed.

Large scale exploitation of a mangrove forest can be carried out sustainably so that the benefits can be maintained indefinitely. The price of sustainability may be a somewhat simplified ecosystem, with lower biodiversity and possibly a decline in productivity, but with the major ecosystem functions intact. A particularly good example, the Matang mangroves of Malaysia, is discussed later in this chapter.

Properly conceived and enforced sustainable management systems are depressingly rare. In too many cases exploitation is completely unregulated, a management system is inadequate or enforcement is lax. The result of overexploitation is habitat degradation, reduction in area and ultimate loss of the resource being exploited. Mangrove loss may also be followed by increased erosion of shorelines, a decline in fish catches or other unforeseen (but hardly unforeseeable) effects.

Unwise exploitation is not the only threat that mangroves currently face. The benefits of mangroves are often undervalued, or simply not recognised, and mangrove habitats are as a result regarded as mere waste land, suitable only for conversion to an alternative use which is seen as more beneficial. Mangroves are the subject of land 'reclamation' for building purposes, or for conversion to agricultural land. (Even though subsequent erosion may reduce the value of the land for building, or a saline and anoxic soil may prove less than ideal for growing crops.) A particularly widespread and serious impact has been the clearance of mangroves for the construction of shrimp ponds.

Conversion of mangrove land to other uses may be justifiable if the gains are greater than the losses, but only rarely have attempts been made to evaluate these fully. Economic analysis of the true value of the products and services supplied by mangroves is likely to prove essential in future rational management. Ecologists may prefer to measure productivity in Joules rather than in dollars, but the latter units have more impact on decision makers.

Mangroves additionally face indirect and accidental threats. Alteration of a hydrological regime by the extraction of river water for irrigation, many miles from the sea, may alter estuarine salinity and sedimentation. This has certainly been a factor in the reduction of mangroves of the Indus Delta in Pakistan (pp. 189–195). Florida offers another probable example of inadvertent damage to coastal mangrove forests caused by changes in terrestrial water use. Pollution, particularly by oil, can also cause severe local damage to mangrove forests. Not all losses are attributable to human actions. Natural hazards, such as typhoons and hurricanes, from time to time cause serious local damage.

Sometimes losses have been mitigated, or even reversed, by habitat restoration and mangrove replanting programmes. Occasionally, just as human activities have inadvertently destroyed mangroves, they may equally unintentionally result in extensions of a mangal. Overall, however, the net effect of the various human impacts has been a drastic overall reduction in mangrove area, although in many countries this is difficult to quantify with any precision. Some figures may highlight the extent of losses in South-east Asia. In the Philippines, around 60 per cent of the original mangrove area has disappeared. In Vietnam, losses are estimated at 37 per cent. In Thailand, 55 per cent was lost between 1961 and 1986 (a year-to-year fall of around 3 per cent). Between 1980 and 1990 Malaysia lost 12 per cent, equivalent to 1.3 per cent annually (Spalding 1997).

Looking beyond the immediate future, mangroves face further threats from predicted changes in climate, due at least in part to human activities: increased atmospheric carbon dioxide concentrations, global warming and concomitant rises in sea level. All of these are likely to have long-term implications for mangrove ecosystems.

An attempt to catalogue all interactions between humans and mangroves would be impractical, as well as tedious. Humans exploit mangrove ecosystems in a multitude of ways, and have many and varied impacts upon them. This chapter therefore discusses only selected examples of some of the major ways in which mangroves and humans interact, over varying time scales (Fig. 2.4).

Uses of mangroves

The benefits of mangroves to mankind vary greatly between areas. Fringe mangroves (p. 34) may contribute relatively little productivity, hence may not be of much significance as a source of wood, or in contributing to fisheries. They are often of great importance in coastal protection. Riverine forests such as those of the Sundarbans of the Ganges delta are extensive in area, generally highly productive and the basis of many products that can be extracted for human use. The precise goods and services available depend on the nature of the forest (Ewel *et al.* 1998).

Direct uses

Accurate information is generally available only for products which are extracted and marketed on a commercial basis. Local subsistence use is less well documented and probably undervalued, and knowledge of it is often little more than anecdotal. It can nevertheless be substantial and important. Some of the uses of mangrove products are listed in Table 8.1.

Mangrove wood is used in construction of buildings, as firewood and in the manufacture of charcoal (p. 175). Branches are used to make fishing poles and fish traps. In some areas, wood extraction is a local industry, managed more or less sustainably in such a way as to benefit local people. In other instances, this is far from being the case. Vast areas of mangrove in South-east Asia have been cleared to support the international woodchip industry, with no attempt at sustainable management (Ong 1995).

Mangrove fruits may be used as food, leaves as fodder for domestic animals (p. 192), and the bark is a source of tannins and in medicine. (Although the medicinal use of mangrove materials may sometimes be overstated. When one group of West Africans was asked about the medicinal uses of mangrove products, they reeled off an impressive list. When asked instead 'what do you do when you are sick?', the answer was 'I go to the clinic').

Table 8.1 Some uses for mangrove products, mainly from FAO (1994)

Fuel	charcoal, firewood
Construction	timber, scaffolds, railway sleepers, mining props, boat building, dock pilings, beams and poles, thatch, matting, fence posts, chipboard
Fishing	fishing stakes, fishing boats, wood for smoking fish, tanning for nets/lines, fish poison, fish-attracting shelters
Textiles	synthetic fibres (rayon), dyes, tannin for preserving leather
Food, drink	sugar, alcohol, cooking oil, vinegar, tea substitute, fermented drinks, dessert topping, seasoning (bark), sweetmeats (propagules), vegetables (fruit, leaves)
Domestic	glue, hairdressing oil, tool handles, musical instruments, rice mortar, toys, matchsticks, incense, cigarette wrappers, cosmetics
Agricultural	fodder
Medical	treatment of: ringworm, mange, toothache, leprosy, sore throat, constipation, dysentery, diarrhoea, boils, bleeding lice, fungal infections, bleeding, fever, catarrh, kidney stone, gonorrhoea, etc.
Miscellaneous	paper manufacture

The most versatile mangrove tree must surely be the nypa palm (*Nypa fruticans*) of Asia, Australia and (following deliberate introduction) West Africa. Nypa leaves are harvested for the construction of dwellings. The mature leaves are dried and the midribs removed; the fronds are then folded round a bamboo rod and stitched in place to make a shingle for roofing or walls. Leaf petioles are used as floats for fish nets, or chopped and boiled to obtain salt. Young leaflets are used as cigarette wrappers, and older ones to weave hats, umbrellas, rain-coats, baskets, mats and bags.

The gelatinous endosperm of young seeds is eaten raw or preserved in syrup, or to flavour ice cream, while the ivory-like hardened endosperm of mature fruits is used to make buttons. Various components are used medicinally, in various parts of South-east Asia, to treat toothache, headaches and herpes.

The sap, which contains 14 to 17 per cent sucrose, is a particularly valuable commodity. It is used to produce a soft drink and, after fermentation, a potent alcoholic beverage. (The medical use of nypa products in the treatment of headaches may therefore seem particularly apposite.) Attempts have been made to develop nypa alcohol production for industrial use. Yield is said to be increased by physical stimulation of the stalk. Methods include regular kicking or beating with a mallet, shaking the fruit to and fro or a complicated routine of bending the stalk 12 times in one direction, patting it backwards and forwards 64 times, kicking its base four times and repeating the whole process four times per week. These elaborate procedures apparently increase the daily sap yield from 155 ml to 1300 ml (Hamilton and Murphy 1988).

Not many mangrove species can boast quite such a wide and varied array of products, but most are of at least local significance in contributing to the economy.

Indirect uses

Mangrove animals are also an important source of food and other products. As with the direct extraction of mangrove wood, leaves and fruit, the nature and extent of these varies from place to place. In many parts of the IndoPacific the larger mangrove-dependent molluscs, such as *Telescopium*, are collected for food or as a source of lime (see Chapter 4). Bivalve molluscs may also be extracted from the mangrove mud. Honey production may depend on bees using man-grove flowers and, of course, various commercial fisheries depend to varying extents on mangroves.

The dependence of fish and shrimp fisheries on mangroves has been discussed in Chapter 6. The extent of the linkage with fisheries is difficult to establish, partly for the reasons already discussed, and partly because only commercial fisheries are reasonably well documented. Subsistence fishing, which does not pass through commercial channels, can be of crucial importance to a local community. A good example of the diversity of mangrove-based fisheries is shown by Sarawak, on the northern coast of Borneo. A wide range of trapping techniques has evolved, to match the diversity of creeks and inlets in the mangal. More than 30 species of fish are caught, about 10 species of shrimp, the mangrove crab (*Scylla* sp.), and two of jellyfish (Pang 1989).

Mangroves and coastal protection

As discussed in Chapter 2, mangrove roots tend to retain sediment, consolidate the soil, hence they facilitate accretion and retard coastal erosion. This ability has been deliberately exploited.

In the Jiulong Jiang estuary, southern China, for instance, there has been a practice for many years of constructing earth embankments across the mouths of inlets and converting the area behind the barrier to shrimp ponds, or rice or lotus fields. In the process, the mangroves that fringed the shores of such inlets were often destroyed. Serious problems arose because the region is affected by typhoons which regularly destroyed the impounding barriers. Replacing the earth embankments with concrete, and even sinking a retired warship as a breakwater, simply showed that man-made materials were no more effective against typhoons than earth embankments. Only when the local villagers were involved in a scheme to plant mangrove trees (*Kandelia*) systematically along the seaward side of earth embankments did they (and shrimp ponds and paddy fields) survive. Mangrove roots compare favourably with concrete and steel as mechanical structures; besides, of course, contributing an exploitable source of wood, and molluscs and crabs for food.

Similar stories could be told of other parts of the world, although it is commoner to find the opposite situation: mangrove removal followed by coastal erosion, rather than mangrove planting resulting in shoreline preservation.

Ecotourism

Mangroves can also be exploited for ecotourism. In Trinidad, tourists are attracted by the scarlet ibises (*Eudocinus ruber*) of Caroni Swamp (p. 68), and in Florida and Honduras by the chance to kayak among the mangroves. The nature reserve of Kuala Selangor, in Malaysia, is a particularly good example of well-organised ecotourism. It lies on the estuary of the river Selangor, within easy reach of the capital, Kuala Lumpur. The reserve area comprises an assortment of pools among the mangroves with pathways and hides from which visitors can watch kingfishers, herons, bee-eaters and numerous other spectacular birds. Walkways have been constructed through the mangroves above high tide level, so that it is possible to observe monitor lizards and crabs in comfort, and without the necessity of even setting foot in the mud. A few kilometres away, the famous synchronously flashing fireflies (see p. 59) can be observed from small boats.

Although ecotourism has not yet been widely developed, it does represent a significant potential source of revenue, particularly where mangrove areas are close to centres of population, or to other tourist attractions.

Sustainable management: the case of the Matang

Very few mangrove forests are pristine: most are to some degree affected by human activities. If exploitation is uncontrolled the result is very often deterioration with possibly loss of biodiversity and frequently reduction in extent of the exploited forest. The consequence is often a drastic reduction of the resource being exploited.

In a few cases, an effective management regime has been put in place, such that intense exploitation is maintained without significant long-term deterioration. One of the best examples is that of the Matang forest of western Peninsular Malaysia, which has been managed on a sustainable basis for close to a century. The managed area of the Matang consists of an estuarine complex of streams, creeks and inlets, amounting to more than 40 000 ha. Some 2000 ha are left untouched as 'Virgin Jungle Reserve', while further patches are set aside for research, or protected as archaeological, ecotourism, educational or bird sanctuary forests.

The principal harvest from the Matang is wood for charcoal, which is a major domestic fuel in the area. The management routine has been modified since its inception, but currently operates on the basis of a 30–year rotation. The forest is divided into blocks of a few hectares. Blocks are allocated to charcoal companies by the Forestry Department, who regulate the whole operation. Each block is clear felled: workers simply move in by boat, demolish every tree with chainsaws and cut the timber into logs of standard length, leaving only a 3–m strip on the shoreward side to prevent erosion of the bank. These logs are ferried to charcoal kilns in a nearby village.

Because the blocks are allocated for clearance in such a way that they are always surrounded by mature forest, repopulation with mangrove propagules occurs rapidly. The debris resulting form the clearing operation takes about 2 years to decompose. After 1 year, the site is inspected. If less than 90 per cent of the area is covered by natural regeneration, repopulation is assisted by artificial planting, mainly with the dominant, and preferred, species *Rhizophora apiculata*. Local villagers are contracted to rear suitable seedlings in small nurseries for this purpose. At this time, weed species can also be removed by hand, or with chemical weed killers. The mangrove fern *Acrostichum* can be a particular problem. It is well adapted to occupying sunlit spaces in the forest, so rapidly latches on to a cleared site, and makes it difficult for mangrove propagules to establish themselves in the soil. Destruction of seedlings by crabs and monkeys (pp. 74 and 90) can be a problem. The following year, the site is again inspected, and any parts where seedling survival has been less than 75 per cent successful are again replanted.

Fifteen or so years later, the site is revisited, and the young trees thinned out to a distance of 1.2 m, using a measuring pole of that length. (In premetric days the thinning distance was set at the round figure of 4 feet, and there has been no reason to change it). The thinnings—all the same age, hence a standard thickness—are valuable as fishing poles. When the stand is 20 years old, it is again thinned, this time to a distance between trees of 1.8 m (6 feet): this time the thinnings (still of uniform thickness) are of a size suitable for the construction of village houses. Because the previous thinning means that the trees are not crowded, these are ideal for their purpose. Finally, after 30 years, the block is again ready for clear felling for charcoal (Fig. 8.1) (Gan 1993; Ong 1995).

The success of the Matang management lies in the allocation of blocks so that no two adjacent blocks are clear felled within a short time of each other. The entire forest is a mosaic of patches of different ages, apart from the Virgin Jungle Reserve and areas set aside for research.

Since management began, there has been a trend towards virtual monoculture of *Rhizophora apiculata* in the intensively managed areas of the Matang. During this time, there is some equivocal evidence of a slight decline in productivity (see p. 169), but overall the Matang is a model of sustainable management of a natural resource—a depressingly rare situation.

In 1992, wood extraction amounted to more than 450 000 t and was worth a little over £2 500 000. In recent years, declining demand for charcoal, and a shortage of workers for the labour-intensive business of timber extraction and charcoal burning suggest that the future value of the Matang may lie in other

Fig. 8.1   A mature stand of mangroves (mainly *Rhizophora*) at the edge of a tidal creek in the Matang, western Peninsular Malaysia. The trees are all virtually the same age and size, a result of the management regime. The understorey vegetation consists largely of the mangrove fern *Acrostichum*.

products. Although the management is largely directed towards timber extraction, this accounts for only around 12 per cent of the total economic value. The area supports thriving fisheries. The offshore waters annually yield more than 50 000 t of fish, valued at £16.9 million, and supporting nearly 2000 people. Farming of the blood cockle (*Anadara granosa*) currently runs at more than 34 000 t year^{-1}, worth £1.8 million, and could be further developed. The Matang also has considerable potential for tourism, being rich in wildlife, including otters, monitor lizards and a wide range of birds such as the rare milky stork (*Mycteria cinerea*). At present, with virtually no infrastructure, tourism probably brings in around £250 000 annually to the local economy (Gopinath and Gabriel 1997).

Even if the market for charcoal disappears, managed mangrove forests such as the Matang should therefore have good long-term prospects, provided the connections between mangrove production and other activities are fully recognised.

Shrimps versus mangroves?

There are many causes of loss of mangroves. Of these, clearance of mangroves for aquaculture, particularly of shrimps, has caused the most controversy. The issues are complex, and the debate between aquaculturists and conservationists often ferocious. Vehemence is no substitute for argument, however. Without getting too embroiled in controversy, this section will examine the relations between shrimp aquaculture and mangrove depletion. Among the issues are the extent to which mangrove loss can actually be attributed to the spread of aquaculture, the overall environmental impact of aquaculture, and whether sustainable mangrove-based aquaculture has been, or can be, achieved.

Aquaculture is not new. Small cultivation of fish and shrimps has been carried out in Asia for many centuries. A good example is shown by the *gei wai* system of southern China. *Gei wai* are shallow ponds formed by impounding areas of mangrove with mud embankments. Deep drainage channels (13 m) are dug round the edge of each pond, surrounding a patch of untouched mangroves, and connected to the sea through a sluice. The growth of shrimp (*Penaeus* and related species) is supported by the mangroves in and around the pond. Often fish and molluscs are cultivated simultaneously in the same ponds. To harvest the shrimp, a net is placed across the sluice, and the pond partly drained at a low spring tide. Shrimp concentrate in the drainage channels; large shrimp are collected, and small ones returned to the pond. As the tide rises, the pond floods again, and the net across the sluice excludes predators, while allowing shrimp larvae to enter to restock the pond. Annual production is not high, less than 500 kg ha^{-1}, but running costs are low. Shrimp larvae enter with the tide and do not need to be caught at sea or reared in hatcheries, while nutrients for shrimp growth are supplied either by mangrove litterfall or by the incoming tides. Most importantly, the process is sustainable, as the mangrove environment is not greatly altered and mangrove productivity is maintained.

Over the last few decades, aquaculture has burgeoned into a 'blue revolution', in response to a combination of increased international demand, the quest for greater profit and concern about over-fishing of natural stocks. Shrimp farming has increased at an annual rate of 20 to 30 per cent, and is now a global industry with an annual production of more than 700 000 t and retail value of over US$20 000 000 (Primavera 1997*b*).

Traditional, small-scale methods, such as the *gei wai* ponds, have given way to more intensive methods of production. The so-called extensive system of rearing is similar to traditional methods, albeit on a larger scale; typical rearing ponds range from 5 to 50 ha. The increased size and intensity of rearing means that ponds cannot be adequately stocked by larvae brought in by the tide. Artificial seeding, usually with larvae caught at sea, is necessary. The food supply is often artificially increased by the addition of fertiliser, and water circulation must be boosted with circulation pumps. Yields increase somewhat, but costs are higher than with traditional methods.

A more fundamental problem is that shrimp depend on mangroves as a nursery area, and as an ultimate productivity base (p. 144). Removal of mangroves to create shrimp ponds undermines the basis of shrimp production. It becomes progressively harder to obtain shrimp larvae to stock the ponds, and the natural food supply brought in by the tides is reduced.

Semi-intensive rearing takes the process further, increasing possible yield by a factor of ten. To achieve this, the stocking rate is greatly increased by using hatchery-reared larvae, water circulation depends heavily on pumping and artificial feeds are required. The increased intensity also means that various chemical additives and antibiotics must be used, further increasing the costs. Intensive systems further increase densities: again, correspondingly greater inputs of food, chemical treatment and water pumping are necessary. Costs can be high, but annual yields may be as great as 20 000 kg ha^{-1}.

Much of the increase in extensive shrimp farming, particularly of the extensive systems, has been at the expense of mangroves. The extent of mangrove clearance specifically due to shrimp farming has been hotly disputed. One estimate is that only 7 to 8 per cent of global mangrove forest loss is due to shrimp aquaculture. Certainly a great deal of mangrove clearance took place before the 1980s boom in shrimp farming got under way. However, even if this figure is correct, it gives a misleading impression of the local impact of clearance for shrimp ponds. In Indonesia, for example, most of the 300 000 ha currently being used to culture shrimps was formerly mangrove forest, and it is planned to increased this figure to more than 1 000 000 ha; Java had lost 70 per cent of its mangroves by 1985, Sulawesi 49 per cent and Sumatra 36 per cent (Macintosh 1996). In the Philippines, approximately half of the mangrove loss between 1951 and 1988 resulted from the construction of shrimp ponds (Primavera 1997*b*). Throughout the world, perhaps 800 000 ha of mangrove have been destroyed specifically for shrimp pond construction.

Mangrove clearance has many secondary effects. In Vietnam, more than two-thirds of the country's mangroves were destroyed by defoliants during the Vietnam War, and development, particularly for shrimp ponds, subsequently continued the process. The environmental consequences have been severe. Coastal erosion has increased. A coastline denuded of mangroves is more vulnerable to storm damage and the inland intrusion of sea water, damaging crops. Fewer shrimp larvae are available to stock aquaculture ponds, and catches of the lucrative mud crab (*Scylla*) have been reduced, as this species also depends on mangroves. When mangrove mud is exposed to the air during pond excavation, it becomes progressively more acid, reducing water quality and shrimp yield. Finally, the increased number of stagnant pools has provided excellent breeding prospects for mosquitoes, and malaria has increased (Macintosh 1996).

Apart from obviously damaging consequences such as these, the more intensive methods of aquaculture leave an extensive 'ecological footprint'. Although there has been a marked increase in the capacity to produce larvae in hatcheries, many ponds are still stocked with larvae caught at sea. The fine-meshed nets used inevitably trap large quantities of other invertebrates and fish larvae as 'by-catch', often amounting to more than 20 times the actual weight of shrimp larvae caught. The impact on other fisheries, and on coastal ecology in general, has not been assessed, but may be considerable.

When pond-reared shrimps are artificially fed, around 30 per cent of the food remains uneaten. The rest—together with a rich brew of chemicals and antibiotics—ends up as effluent. If not excessively concentrated, this waste—rich in nitrogen and phosphorus—represents nutrient which can potentially stimulate mangrove productivity. Depending on the precise method of culture, a 1–ha shrimp pond requires between 2 and 22 ha of neighbouring mangrove forest to assimilate its effluent. Taking into account the other ways in which a mangrove ecosystem can support shrimp pond functions, an alternative estimate is that each hectare of semi-intensive shrimp farm requires 35 to 190 ha of adjoining ecosystem, mainly mangrove (Kautsky *et al.* 1997; Robertson and Phillips 1995). If effluent production is greater than can be readily assimilated by surviving mangroves, its ecological impact can be disastrous.

Shrimp farming, then, is not without its problems. In several countries, the 'blue revolution' reached Klondyke proportions, economic and ecological problems became apparent, and an unsustainable industry crashed. Taiwan is a good example. In the late 1960s, shrimps became the dominant aquaculture crop. By 1980, around 2000 ha of ponds were producing 5000 t of shrimp. Hatchery production of larvae rose from 6000 in 1968 to 300 000 000 in 1979, to 1.3 billion in 1984 and 3 billion in 1985. By 1987, annual shrimp production reached 100 000 t. By the mid 1990s the industry had virtually collapsed. Overstocking, overcropping, misuse of processed feeds and antibiotics, problems with effluent disposal and a variety of bacterial and viral diseases caused a fall in production to 30 000 t in 1989. Similar booms followed by slumps have occurred in China, Thailand and other shrimp-producing countries (Baird and Quarto 1994; Primavera 1997*b*).

When a shrimp industry collapses, what is left is often a devastated coast, with derelict ponds in place of a previously extensive mangrove habitat. These can sometimes be rehabilitated, but the process takes care, time and investment resources which are often not available.

The often disastrous history of shrimp farming in mangrove areas does not mean, of course, that broadly sustainable cultivation of shrimp is impossible. It is clear that constructing shrimp ponds actually at the expense of mangrove is not sensible. In fact much of recent shrimp pond construction has actually been on the landward side of the mangrove belt, above the high tide mark. If the pond area is not too great in relation to the surviving area of mangrove habitat, and if water circulation is appropriately designed, it would even be possible for mangroves to be used as an efficient processing system to limit damaging effects of effluent on the environment. With hatchery-reared larvae to stock the ponds, the destructive trapping of larvae at sea would be largely unnecessary.

Most of the serious problems of shrimp farming could, in principle, be solved: sustainable aquaculture, without massive destruction of mangroves, is possible.

Mangroves and pollution

Pollution comes in many forms, including exposure to hot water outflows, to toxic heavy metals, pesticides, sewage or oil spills. Sometimes the contamination is accidental; sometimes deliberate, because mangroves are seen as valueless, fit only for dumping unwanted wastes.

Thermal pollution comes in the form of water from power plant cooling systems. When *Rhizophora mangle* is exposed to a 5°C rise in water temperature, it responds by decreasing leaf area and increasing the density of aerial roots. Seedlings are more vulnerable than adults to high temperatures (as to low temperatures), and a rise of 7 to 9°C is sufficient to cause 100 per cent mortality (Lugo and Snedaker 1974; Pernetta 1993). High temperatures also greatly reduce the species richness of the mangrove fauna.

Thermal pollution of mangrove habitats is relatively rare. Various forms of chemical pollution are more frequent, the most serious being heavy metals, pesticides, sewage and petroleum products. Mangroves often face combinations of these pollutants.

Heavy metal contamination comes from mine tailings and industrial waste, and includes in particular mercury, lead, cadmium, zinc and copper. Because assay of heavy metals is simple, there is a great deal of information about their distribution and accumulation within mangrove habitats, and rather less understanding of their biological effects. Much accumulates in mangrove sediments, where it may not be in a position to have profound ecological effects. Mangrove trees themselves may be relatively immune to the toxic effects of heavy metals, but the mangrove fauna may be more vulnerable. Mercury, cadmium and zinc are acutely toxic to crab larvae and heavy metals in general cause physiological stress and reduced reproductive success, even at sublethal concentrations. Moreover, accumulation in fish, shrimps or edible molluscs could present serious health problems for the human population (Ellison and Farnsworth 1996; Peters *et al.* 1997).

Much the same can be said of herbicides and pesticides, which usually enter the mangal in run-off from agricultural land. These may also have both acute and chronic effects on both mangrove animals and plants, and can accumulate in the tissues of food species. However, many of the compounds are strongly absorbed onto sediments, and readily degraded in anaerobic soils (Clough *et al.* 1983).

The impact of sewage pollution on mangroves depends very much on the amounts involved. Mangrove growth and productivity may be limited by the available nitrogen and phosphorus, as well as by high salinity (p. 16). Sewage is rich in both nutrients, and is of low salinity. Up to a certain level it would therefore be likely to enhance mangrove productivity; indeed, field trials have shown that mangroves can provide a useful method of waste water treatment. High nutrient loads stimulate excessive algal growth and cause deoxygenation of the water by microbial activity. Algae may grow profusely on pneumatophores and aerial roots, impeding gas exchange, or drifts of the sheet-like green alga *Ulva* smother seedlings and may even physically dislodge them from the soil.

The most obvious damage to mangroves, however, comes from oil pollution. Between January 1974 and June 1990 there were, in the tropics, at least 157 major oil spills (greater than 1000 barrels, or 250 000 litres) from ships and barges. More than half of these were close enough to the coast to present major threats to coastal ecosystems such as mangroves (Burns *et al.* 1993). In addition, there were numerous small spillages involving smaller volumes, often occurring during loading or discharging of tankers, or through illicit washing out of tanks at sea.

Fuel oil consists of a complex mixture of components which differ in molecular weight and in volatility. Lighter fractions are generally more toxic, but often evaporate before an oil slick reaches the shore or soon afterwards. The residual heavy components often kill mangroves by coating pneumatophores and aerial roots, clogging lenticels and killing roots by asphyxiation. The immediate result is often defoliation of the affected trees; depending on the level and persistence of the pollution, trees either recover or die.

The extent of the damage obviously depends on the amount of oil reaching the shore, which in turn depends on wave and wind conditions. The tidal regime also makes a difference: in the Caribbean, where the normal range is often only 30 to 50 cm, narrower zones are affected than in parts of the world with more typical 2 to 3 m tidal ranges. Shore topography and environmental setting also matter: the area of mangroves affected by a given volume of oil will differ between steep and shallow shores, between straight and convoluted coastlines and between fringing, estuarine and overwash mangals (Chapter 2).

If a mangrove forest is heavily blanketed with oil, and suffers locally cata-strophic effects on fauna and flora, it is not hard to link cause and effect. It is not so easy to evaluate subtler, long-term effects on tree health and productivity, or on community structure, without a thorough understanding of the situation before the impact. A further problem in assessing the consequences of oil pollution is that oil spills are unpredictable and there is rarely any baseline environmental information for an adequate 'before-and-after' comparison.

An exception to this is Bahía las Minas, on the Caribbean coast of Panama. This area, fortuitously, includes the Galeta Marine Laboratory resulting in much background information about the condition of the mangroves before the pollution events. The first of these was in December 1968, when the tanker *Witwater* foundered in rough seas, releasing between 2.8 and 3.8 million litres of fuel oil. Because of the time of year, tidal range and rainfall were both relatively high. This combination of circumstances helped to keep channels and larger streams relatively clear of oil. About 49 ha of mangroves were killed, 4 per cent of the mangroves in the bay. By 1979, most of the deforested areas had been repopulated, with the exception of around 3 ha where the sea had succeeded in encroaching on the denuded shore.

The second major spill came in April 1986, when the rupture of a storage tank at a nearby refinery released around 8 million litres of crude oil. At first, this was concentrated in an embayment, where it deposited as a layer 5 mm thick. Approximately half of the trees in this area died within 2 months, leaving few survivors in the lower half of the shore. A few days after the spill, winds and rain combined to wash an extensive slick out to sea. This was later washed ashore over a wide stretch of coastline. Because of low tidal range at the time, the effects were concentrated in a relatively narrow band bordering the shoreline. Overall, an estimated 69 ha of mangrove trees died. There was also massive mortality of the intertidal fauna, with sessile animals suffering more rapid, as well as long-term, effects than mobile species. Aerial roots of *Rhizophora* were reduced by between one-third and three-quarters, reducing the area available for epifaunal communities (Ellison and Farnsworth 1996).

Acute effects were obvious in the immediate aftermath of the 1986 spill. Some of the deforested area was rehabilitated by artificial replanting, although again a few hectares were irrevocably lost by encroachment of the sea. When the site was assessed several years after the 1986 spill, large areas of low density, open canopy remained, suggesting that natural regeneration was slowed by long-term sublethal effects of the oil. One factor delaying regeneration was probably the unusually high mortality of seedlings. Taking this into account, the area affected by the 1986 spill was approximately 377 ha, or 42 per cent of all mangroves in the bay. The total area affected was therefore five to six times greater than that completely cleared (Duke *et al.* 1997).

Some of the oil was absorbed into the sediment, and released intermittently over the following years, so that the initial spill cast a long shadow of chronic ecosystem effects. Five years after the 1986 event, secondary oiling was still taking place, and oil levels in bivalve molluscs were still high. A reasonable estimate is that a mangrove forest may take 20 to 30 years to recover fully from a major oil spill (Table 8.2). As has already been shown in Bahía las Minas, a mangrove forest may be hit by a second oil spill before it has recovered from the first.

Hurricanes and typhoons

Not all destructive impacts on mangroves are caused by man. Tropical coast-lines on the western side of the major oceans are frequently struck by cyclonic storms. In the Caribbean and Gulf of Mexico, these are termed hurricanes; and in the western Pacific, typhoons. In either case they are characterised by extremely high wind speeds, torrential rainfall and storm surges several metres above the normal sea level. In season, there may be several typhoons or hurricanes a month, with particularly serious ones every few years. Vietnam, for instance, has 8 to 10 typhoon strikes each year, with winds of more than 100 km h^{-1} and tides 2 to 3 m higher than normal. Heavy rains may some-times increase sediment movement from the land, accelerate the rate of accre-tion of a shore, and enable mangrove expansion. In most circumstances, however, the impact of typhoons is highly destructive. The Pacific is also affected by tsunamis, 'tidal' waves generated by undersea earthquakes or vol-canic activity. A tsunami may travel at up to 1000 km h^{-1}, and hit the coast as a wave 50 m high.

Table 8.2 Generalised response of mangrove forests to oil spills (compare with Fig. 2.4), from Lewis (1983)

Stage	Impact
Acute	
0–15 days	Deaths of birds, fish, invertebrates
15–30 days	Defoliation and death of small (<1 m) mangroves
	Loss of aerial root community
Chronic	
30 days–1 year	Defoliation and death of medium (<3 m) mangroves
	Tissue damage to aerial roots
1–5 years	Death of larger mangroves
	Loss of oiled aerial roots
	New aerial roots deformed?
1–10 years	Reduced litterfall
	Reduced reproduction
	Reduced survival of seedlings
	Reduced growth or death of recolonising saplings?
	Increased insect damage?
10–50 years?	Complete recovery of forest?

Mangrove forests can absorb much of the energy of the average cyclone, but a severe hurricane can be devastating. One example is Hurricane Andrew, which passed across the south-western coast of Florida in August 1992, at wind speeds greater than 240 km h^{-1}, accompanied by a 5–m storm surge. Heavy damage occurred to about 150 km^2 of mangroves. About 60 per cent of the trees, particularly the larger ones, were either uprooted or broken; of the upright and unbroken trees 25 per cent were dead and 86 per cent defoliated to varying degrees. Many of the survivors subsequently died. Initially about 20 per cent of large *Avicennia* trees died, but this rose to 50 per cent, or even higher, in the year following the hurricane (McCoy *et al.* 1996).

Such a major disturbance is likely to have significant, long-term effects on the ecosystem. Of the three major species in the forest, *Rhizophora mangle* survived better than *Avicennia germinans*, which in turn fared better than *Laguncularia racemosa*. Selective destruction therefore had a noticeable effect on the species composition of the surviving trees.

The damaged plots were rapidly recolonised by seedlings. In many heavily damaged plots, recolonisation was mainly by *Laguncularia*, in others by *Rhizophora*, with local soil conditions presumably dictating which species predominated. Only a few *Avicennia* seedlings appeared. Subsequent seedling mortality was high, and growth slow, particularly in the most disturbed areas. This indicated long-lasting effects of Hurricane Andrew on the mangrove community, with slow recovery and a significant change in community structure. However, hurricanes are not the only events affecting the Florida mangrove community. In January 1997, unusually low temperatures selectively killed smaller *Laguncularia* and *Rhizophora*, while the more cold-tolerant *Avicennia* survived.

A detailed model has been developed of the interactions between the three mangrove species of Florida following hurricane damage. This takes account of recruitment, growth and survival rates and how these vary in response to different conditions of salinity, and nutrient and light availability. The model predicted that an open space within the forest canopy would be dominated initially by *Laguncularia* seedlings, but that over time this species would be replaced by others, culminating in dominance by *Avicennia*. This is exactly what happened following Hurricane Andrew (and, previously, Hurricane Donna in 1960). Dominance by a single species is inhibited, apparently, by the intermittent impacts of hurricanes which create gaps in which the successional process can begin again (Chen and Twilley 1998).

Small trees survive hurricanes better than large. Forests in the hurricane zone, such as those of Florida and Puerto Rico, show a fairly uniform tree height, with no exceptionally large trees emerging from the canopy. In Panama—outside the hurricane zone—emergent trees are common, and total forest biomass can be twice that of comparable forests in Florida or Puerto Rico (Lugo and Snedaker 1974).

Intermittent meteorological events are natural 'experiments' on a scale that an ecologist would shrink from contemplating, and reveal much about the interaction between the different tree species and their environment. Small trees survive hurricanes better than large, so that small size and early maturity are promoted. Some species are favoured at the expense of others: *Rhizophora* is more resilient when exposed to high winds, *Laguncularia* is a good coloniser and *Avicennia* the most cold-tolerant. Such irregular events can therefore be important in shaping community structure.

Mangrove rehabilitation

Although throughout the world mangrove habitats are being drastically reduced in area, there are some exceptions to the trend. In parts of New Zealand and Australia, for instance, *Avicennia* patches are spontaneously expanding. Although the reasons are not clear, the expansion is probably a response to inadvertent alteration of sedimentation and other features of the environment by human activities.

In many countries great efforts are being made to restore previously destroyed mangrove habitats by replanting programmes, or even to plant mangroves where none were known before. The reasons are various. In some cases, the purpose is to conserve or recreate an ecosystem for its own sake. More commonly, replanting is carried out because of an awareness of the value of the mangrove resource for fisheries or other activities, or to curb coastal erosion.

Mangroves which produce large propagules, such as *Rhizophora*, are good candidates for replanting, since in principle little needs to be done besides removing propagules from one site and inserting them into the mud at another. In practice, matters are rather more complicated. Newly-established propagules often suffer heavy mortality from crab or monkey attack. Death can result from toppling by oysters, barnacles, algae or other organisms settling on the seedlings. These problems are largely overcome by growing the propagules in nurseries for a few months and planting them out when large and robust enough to cope. Sometimes, though, fouling organisms must still be removed by hand.

Choice of a replanting site is critical. Appropriate conditions of soil texture, salinity and other physical variables are essential. If mangroves have disappeared from a certain location because human activity has rendered it untenable, attempts at replanting are unlikely to be successful. Despite these limitations, some tens of thousands of hectares have been successfully replanted, for instance in the United States, the Philippines, Ecuador, Kuwait and Pakistan. The Pakistan replanting programme is discussed further in a later section (p. 193).

Mangroves and global climate change

It is now accepted that the world's climate is changing. In particular, the concentration of carbon dioxide in the atmosphere has increased, and mean global temperature and sea level have both risen, and are likely to continue to do so. There is controversy over the extent of past and probable future change. There is also disagreement (often fierce) over the precise causes, and particularly over the contribution of human activities to climatic change. For present purposes, however, these controversies need not concern us. Whatever the causes, what is the nature of climatic change, and how might mangroves be affected?

Rise in atmospheric carbon dioxide

Measurements of atmospheric CO_2 indicate a rise from around 280 parts per million (ppm) in the late 18th century to 355 ppm in 1991. Currently the level continues to rise at around 0.5 per cent, or 1.8 ppm, each year. Much of the rise is due to human activities, notably the burning of fossil fuels. Although there are plans to reduce emissions, CO_2 will continue to increase for the foreseeable future (Graves and Reavey 1996).

Plants depend on CO_2 for photosynthesis: they might therefore be expected to respond to a rise in atmospheric CO_2 levels with increased photosynthesis and growth. This has been demonstrated in some, but not all, mangrove species tested. CO_2 assimilation interacts in complex ways with other aspects of mangrove physiology. The stomata through which CO_2 enters a leaf for photosynthesis are also the main route of water loss in transpiration. At higher atmospheric CO_2 levels, stomatal conductance is reduced and water loss falls while CO_2 uptake levels are maintained. The result is an increase in water use efficiency (p. 16). The trade-off between water use and CO_2 acquisition means that the mangrove response to high atmospheric CO_2 may combine increased water use efficiency with varying effects on transpiration rate and growth, depending on other circumstances.

Elevated CO_2 similarly causes greater efficiency of nitrogen use. In general, a rise in atmospheric CO_2 can be expected to stimulate growth when it is limited by water, carbon or nitrogen, but not when salinity is too high for water uptake to be maintained, or when some other nutrient is limiting (Ball and Munns 1992; Field 1995).

Short-term experiments are of limited value in predicting long-term trends. When plants are exposed to elevated CO_2 for a long period, the initial increase in photosynthesis is not maintained; compensating mechanisms tend to bring the photosynthetic rate back towards its previous level. Over still longer periods, alterations in plant morphology may also reduce the impact of elevated CO_2. Comparison of herbarium specimens has shown a decrease in the density of stomata (admittedly not in mangrove species) since the Industrial Revolution, which correlates closely with the increase in atmospheric CO_2. With higher prevailing levels of CO_2 in the atmosphere, a given level of photosynthesis can be maintained with fewer stomata (Graves and Reavey 1996).

To the extent that different species of mangrove respond to different extents, community composition might be altered. On balance, however, it seems unlikely that the direct effects of rising CO_2 will be as important as the other aspects of climate change. The same is probably true of the mangrove fauna.

Global warming

Since about the mid 18th century, the world has been warming. This may be due to long-term trends, related for example to variation in eccentricity of the Earth's orbit round the Sun. The temperature trend does, however, correlate with the increase in atmospheric carbon dioxide (see last section). This may contribute to an enhanced 'greenhouse' effect.

The Earth absorbs solar radiation, predominantly at shorter wavelengths. In turn, the Earth emits radiation: this is largely at longer wavelengths, in the red and infrared reaches of the spectrum. Much of this is absorbed by the so-called 'greenhouse gases' in the Earth's atmosphere. Were it not for this trapping of energy the Earth's surface would be a great deal cooler. The most important natural greenhouse gas in the atmosphere is water vapour, followed in importance by carbon dioxide and methane.

Human activities have caused atmospheric CO_2 to rise (see last section). Methane is also rising, by around 0.9 per cent annually. The main source is anaerobic respiration in the guts of domestic animals and in swamps and other waterlogged soils, together with a contribution from mining and oil and gas extraction. The greenhouse effect is further enhanced by relatively small amounts of chlorofluorocarbons (CFCs), which are particularly potent greenhouse gases.

There has been an increase in global mean temperature of between 0.3°C and 0.6°C since the late 19th century. Depending on a number of assumptions, it is predicted that temperature will increase by 0.2°C to 0.5°C per decade during the next century. By the year 2025, global mean temperature would therefore be about 1°C above its present value, and 3°C by the end of the 21st century (with smaller increases in the tropics). Accompanying this will be a substantially wetter atmosphere and generally (but not universally) increased rainfall. It is possible that tropical storms will be more frequent and more severe (Field 1995).

Mangrove species differ in their responses to raised temperature. Photosynthesis, and other variables such as leaf production, usually show clear temperature optima, falling off at lower and higher temperatures. The optimal temperature varies between species and between localities. Among Australian species, most peak between 21°C and 28°C. *Avicennia marina*, which is found furthest south, has a peak at 20°C, while the more tropically-distributed *Xylocarpus* peaks at temperatures higher than 28°C (Hutchings and Saenger 1987). In some plant species, the optimal temperature for photosynthesis has been shown to alter in individual plants with seasonal temperature changes. It is not known how far mangroves can adapt in this way.

Given the range of temperatures that mangroves experience in their daily lives—which may amount to more than 20°C in temperate species—it seems unlikely that the rises predicted will make a great deal of difference to mangrove productivity. The geographical distribution, however, may be affected. As discussed already (p. 32), the limits in the latitudinal distribution of mangroves correlate closely with temperature. An increase on mean global temperature would, therefore, tend to allow northward and southward extension of the geographical distribution of mangroves. This spread would be limited by topography: within the India Ocean, for example, there is no land to the south of South Africa, and no sea north of the Red Sea and Arabian Gulf. On the shores of the Pacific and Atlantic, however, a modest expansion of the total range of mangroves might be expected, probably at the expense of salt marsh habitats. Because species vary in their temperature tolerance, species would presumably spread to different extents and the combination of mangrove species at different locations might alter.

Raised temperature would have some effect on other aspects of the ecology of mangrove habitats, such as the acceleration of litter decomposition, and effects on the physiology and geographical distribution of components of the mangrove fauna. These effects are not likely to be great and are, in our present state of knowledge, unpredictable.

Sea level rise

One of the effects of global warming is a rise in overall (eustatic) sea level. This stems largely from thermal expansion of the world's oceans, with a smaller contribution from the melting of glaciers and major continental ice caps. The best, current estimate suggests that global sea level is rising at around 6 cm per decade. This predicts that by the year 2030 mean sea level might have risen 18 cm above its present level (with a range of 8 to 29 cm); the expected rise is 44 cm (21 to 71 cm) by the year 2070, and 60 cm by the end of the 21st century (Field 1995).

The land also rises and falls, due to tectonic activity, to isostatic readjustment of the Earth's crust following the post-ice age redistribution of ice and water, and to the balance between the rates of sediment deposition, erosion and compaction. Eustatic sea level rise will not be experienced uniformly around the world, but may translate locally into a greater or lesser rise, or even a fall.

Much also depends on the environmental setting of the mangroves (Chapter 2). In an estuary, mangrove trees (and mangrove animals) often show a particular pattern of species zonation up and down the river, which is dependent largely on a salinity gradient, which is the outcome of the interplay of river and tidal flows. A rise in sea level would have the effect of shifting the zonation pattern in an upriver direction (offset by any increase in local rainfall which might increase the river flow). Upshoredownshore zonation patterns might also move, in a landward direction. Trees on the seaward fringe would be inhibited by extended submergence times, while those on the landward margins might be able to extend, provided suitable habitat was available (Pernetta 1993).

If neither of these situations applies, the outcome may depend on whether the rate of deposition of sediment can keep pace with the rate at which sea level is rising. In the great river deltas, on high oceanic islands with considerable runoff from rainfall or in situations where large amounts of marine sediments can be trapped, mangroves may avoid losing the race. On small carbonate-based islands, or arid zone mangroves such as those of the Red Sea, this is unlikely to be the case.

Past changes in sea level, and ecological changes associated with them, can give valuable clues to probable future changes. In northern Australia, a dramatic rise in sea level between 8000 and 6000 years ago flooded the coastal plains. In some cases deposition of sediment (mainly originating in the sea) kept pace with the rising sea level and extensive mangrove forests established themselves. When the sea level stabilised, further sediment deposition allowed these to expand seawards (Wolanski and Chappell 1996). Sea level rises can create opportunities as well as destroy them. An alternative guide to future prospects is the study of changes known to have occurred in the Caribbean and Pacific regions during the Holocene period. The stratigraphic record of mangroves during this time of suggests that mangrove ecosystems in these regions could keep pace with a eustatic sea level rise of 8 to 9 cm per 100 years. Between 9 and 12 cm per 100 years they are under stress; above this rate of rise the outcome is likely to be complete collapse of the ecosystem (although individual trees could survive). Faced with future rises on the scale predicted, many island mangrove habitats could not persist (Ellison and Stoddart 1991).

Prediction, where many of the key variables to take into account are poorly understood, is a hazardous business. Models of the world's climate and ocean current systems are still not sufficiently sophisticated to give a reliable picture of the immediate future. Forecasts of rainfall, temperature and sea level a century hence are even less accurate. If the general picture is correct, most mangrove communities will be affected by global climate change. Some will adapt with minor adjustments of zonation and species composition. In other areas, mangrove species composition will change as geographical ranges expand, and some temperate salt marsh habitats might be replaced by mangroves. Where mangroves are hemmed in by unsuitable terrain, sea level rise is likely to cause compression of the mangrove zone. Finally, in places where sedimentation rates are low, mangroves might simply disappear.

Mangroves of the Indus Delta: a case study

The mangroves of the Indus delta, Pakistan, exemplify almost all of the issues discussed earlier in this chapter, and therefore make a useful case study (Hogarth 1999).

Without the River Indus, Pakistan would be virtually a desert. The river winds its way southwards through arid plains, to debouch into the Arabian Sea through a complex delta of tidal creeks and inlets. The bulk of Pakistan's mangroves are found in this delta, the largest area of arid zone mangroves in the world. Annual rainfall is negligible, often below 200 mm (much of it falling in a single day!), and less than the evaporation rate. River water is the main source of fresh water, and large parts of the delta which do not receive significant river flow actually become more saline than the sea itself.

Rivers in alluvial plains are generally unstable, and the Indus has been no exception. Silt builds up, raising the river above the surrounding plain but constraining it within banks of sediment. Periodically, these banks burst, flooding the adjoining plain. The river then establishes a new bed, and the process is repeated. Alluvial deltas are also unstable: channels become blocked with sediment and mud banks are eroded by wave action, so that the principal distributary channels shift around.

In the case of the Indus, geological events have also changed the course of some of the delta channels. The result is that the outflow has varied between more than a dozen separate channels, and the principal outflow has ranged between Karachi and the Indian border (Fig. 8.2). A river that writhes around in this way proved difficult to accommodate within a stable pattern of human settlement. Since the 19th century, the Indus has therefore been progressively tamed by the construction of retaining embankments, so that the area now receiving direct river outflow is confined to an 'Active Delta' of only 119 000 ha.

A further trend has been to divert more and more of the Indus water for irrigation, and for industrial and domestic water supplies. Irrigation engineers measure water volume in a unit known as the MAF, or million acre feet. One MAF is the volume of water that would cover an area of a million acres 1 foot deep, equal to 1 230 000 000 m^3. The MAF, although not metric, at least has the advantage of being on a scale appropriate to the Indus. In the past, annual flow down the Indus has been around 150 MAF; roughly enough water to cover an area the size of the mainland United Kingdom more than half a metre deep. The Indus is—or was—a big river.

Since the construction of the latest of the major dams in 1955, the annual flow has sometimes been as low as 30 MAF, and in the dry season virtually no water reaches the coast through the Active Delta. (Fig. 8.3). Near Karachi, some water arrives through two small rivers, rich in effluent of various sorts, but at least of low salinity. Elsewhere, virtually the only fresh water arrives from drainage of irrigated land. Reduction of freshwater flow leads to the intrusion of sea water further and further into the delta. The mangroves (and agricultural land within the delta) are exposed to increasing salinity.

Reduced river flow also reduces the amount of sediment delivered, from around 200 000 000 t in the past to less than a quarter of this amount in a typical year. Reduction in sediment delivery affects the equilibrium between deposition and

Fig. 8.2 The Indus Delta, indicating the distribution of mangrove vegetation inferred from the interpretation of satellite imagery. The arrows indicate the current mouths of the Indus. (Reproduced with permission of IUCN-Karachi.)

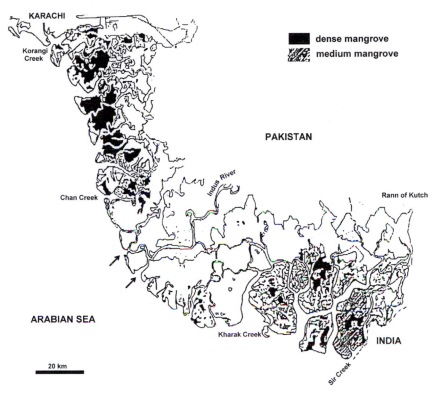

Fig. 8.3 Flow, in MAF (see text for explanation), down the lower Indus from 1940 to 1994. The upper and lower bars indicate flow during the wet and dry seasons, April–September and October–March, respectively. (Data from Milliman *et al.* (1984) and WAPDA, Pakistan.)

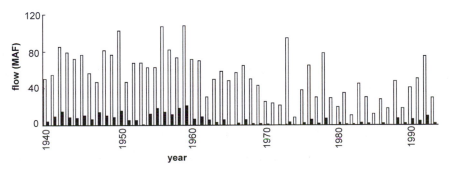

erosion that is a feature of any sandy or muddy coast, causing regression of the shoreline. Reduction in sediment makes it less likely that mangroves will be able to keep pace with rising sea level, which is forecast at 6 mm year^{-1} in the Indus delta (Farah and Meynell 1992).

What effects will these changes have on the Indus delta mangroves? To assess this, it is necessary first to have a clear idea of the current status of the mangroves, and how they have changed in the recent past. Satellite imagery is a useful source of information. Different surfaces reflect different wavelengths of light, so that remote sensing techniques can be used to distinguish areas covered by vegetation from those that are bare mud. In principle, it should therefore be a straightforward matter to map the distribution of mangrove vegetation, measure its area and detect changes with time. In practice, there are serious complications. Different analyses use different methods to identify and categorise mangrove areas. Only rarely is crucial information given about the criteria used, and even more rarely is the analysis 'ground-truthed' by systematically visiting sites and comparing what is visible on the ground with what was inferred from satellite photographs. In some cases algae covering the mud have probably been misinterpreted as mangroves. These, and other complications, explain why estimates of total area sometimes diverge widely.

The best estimate is that the Indus delta currently contains rather less than 200 000 ha of mangrove, and that this area is declining, possibly by as much as 2 per cent annually. What satellite images do not reveal is that a significant proportion of this consists of trees stunted by their stressful environment, growing no more than half a metre in height.

In addition to the reduction in area, it appears that diversity has been reduced. Historical records and pollen deposited in the soil suggest that in the past nine species of mangrove occurred within the delta (although there is some doubt about the correct identification of two of these). At present, natural mangal consists almost entirely of *Avicennia marina*, with small and localised patches of only two other species, *Ceriops* and *Aegiceras*.

Is the decline in mangrove area and diversity due solely to deprivation of fresh water? This cannot be the whole story. If lack of fresh water was the only factor, mangroves should survive better in the area still served directly by the Indus. This is not the case: the Active Delta is in fact virtually devoid of mangroves (Fig. 8.2). Why should this be? The areas on either side of the Active Delta are controlled by the Forestry Department of the provincial government. Control is imperfect, but does go some way towards limiting the use of mangroves for firewood and as grazing for domestic animals. This is not the case in the Active Delta, where virtually no regulation exists. Around 26 000 camels, sheep, goats, water buffalo and cattle are believed to spend at least part of the year in the area. These feed on *Avicennia* (Fig. 3.7), demolishing branches and trampling seedlings, and compacting the soil to the extent that it becomes virtually impossible for propagules to establish themselves. Overgrazing, the result of poor management, is a major factor in the destruction of the mangroves.

In other parts of the delta, mangroves have also suffered, but survived. Near Karachi, they have been cleared for harbour and other developments, or been buried in spoil from dredging. Overexploitation has opened up cleared areas round many of the villages. Oil, heavy metals, pesticide residues, domestic sewage, animal waste and assorted industrial effluents enter the creeks round Karachi by hundreds of millions of gallons daily. Only a few kilometres from the outfalls, mangrove trees seem to flourish.

What are the probable consequences of the decline in Indus delta mangroves? At least 100 000 people in the delta depend on mangroves for firewood and fodder, and any further decline in mangroves would affect them adversely. Apart from their exploitation for firewood and fodder, mangroves serve to limit coastal erosion. Pakistan's second largest harbour, Port Qasim, is protected by a mangrove-covered mud bank. Outside this natural harbour defence waves reach a height of 6 m; inside, the maximum height is 0.5 m. Without such coastal defences it is hard to see how Port Qasim could continue to function.

There are also likely to be less local impacts of mangrove loss. Many thousands of birds over-winter in the delta, having migrated down the Indus valley from Central Asia. Reduction of crucial winter feeding grounds would have ecological effects many thousands of kilometres away.

As in other parts of the world, the Indus mangroves sustain offshore fisheries. The shrimp fishery, mainly for export, is worth around $60 000 000 annually, and other fisheries are probably worth a further $30 000 000 or so. One useful measure of the state of a fishery is the catch per unit effort (CPUE). Figure 8.4 shows the change in CPUE for fish caught before and after completion of the last major dam (so far), in 1955. The shrimp fishery shows a similar pattern. Other factors, such as over fishing, contribute to a declining CPUE, but CPUE fell sharply before expansion of the fishing fleet. Completion of the dam and reduction of river flow are therefore likely to have been significant factors. It is not possible to assess the relative effects of declining mangroves, and of the reduction in fresh water and sediment.

Efforts are being made to reduce or even reverse the process of decline by a replanting programme. The aims were discussed with local villagers at all stages: without local understanding and involvement, replanting projects have little chance of success.

Initially, *Rhizophora mucronata* propagules were collected from elsewhere in Pakistan and grown in nurseries for up to a year before being planted out on selected and carefully prepared sites. During their first years of life, the seedlings are given intensive care, with the removal (by hand) of algae, propping up of any toppled seedlings, and judicious pruning to encourage vigorous growth of a main stem (Fig. 8.5). Given this assistance, the survival rate is better than 95 per cent, and after 3 years or so the planted mangroves reproduce for the first time. Although it is too early to evaluate the long-term success of the project, around 12 000 ha have now been replanted (Qureshi 1996). The mangrove replanting

Fig. 8.4 Yearly variations in total fish catch (x 1000 tonnes), and in catch per unit effort (CPUE) (tonnes per boat) along the coast of Sindh between 1949 and 1980. Construction of the Kotri barrage was in 1955. (Data from Milliman *et al.* 1984.)

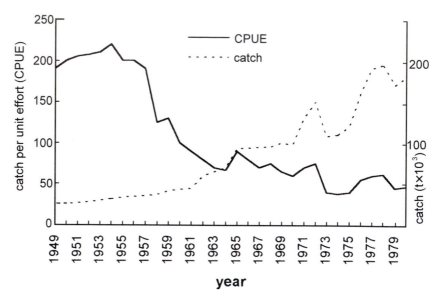

Fig. 8.5 Young *Rhizophora apiculata* planted out in a denuded area of the Indus delta near Karachi, Pakistan. To increase tidal inundation frequency, the seedlings have been planted along a shallow irrigation ditch.

nurseries have been made a feature of a small ecotourism project in the area, to help raise awareness of the significance of mangroves (Fig. 8.6).

What are mangroves worth?

While it is obvious that mangroves are of economic value, putting figures to that value is not straightforward. The total economic value of a mangrove area must take account of the whole range of goods and services provided, which differ from one mangrove to the next. Many of these cannot be measured directly (Ewel *et al.* 1998; Spaninks and van Beukering 1997).

Marketable commodities generally can be measured because they have an identifiable volume and cash value. Charcoal production in the Matang forest is a good example (p. 175). Prices vary with demand and (with exported mangrove products such as shrimp) with exchange rates, but in the short term such fluctuations are generally not a great proportion of the total.

Fig. 8.6 Walkway constructed as part of an ecotourism project among the mangroves of the Indua Delta, close to Karachi. The mangroves on either side of the walkway are not seedlings but adult trees, stunted in their growth by environmental factors such as hypersalinity, pollution and soil compaction.

Generally these are quoted as gross financial benefits, without taking collecting costs into account.

Products which are not marketed are more difficult to assess: local villagers collect firewood or timber, or catch crabs and fish, for their own use, without money changing hands. These, too, can be estimated, and economic values assigned, often on the basis of the market price that would have been paid for such products had they been purchased.

Problems arise with indirect products whose availability depends on mangroves, but which are harvested at a distance, such as offshore fisheries. Assigning economic values of mangroves to offshore fisheries means making assumptions about the nature of the linkage between the two: as already discussed (p. 144), the relationship is elusive. In practice, economists usually tacitly accept an established correlation between mangrove area and fish yields as implying dependence.

The contribution of mangroves to aquaculture yields is also problematical. Traditional *gei wai* aquaculture (p. 177) is very largely mangrove dependent, as is the aquaculture of crabs and molluscs, common in Malaysia, which is based on mesh cages floating in mangrove creeks. Neither technique is destructive of mangroves. Where farmed shrimps are fed artificially, they do not depend on mangroves as an energy source, but may do so as a means of effluent disposal. In this case, too, mangroves supply an essential service in maintaining aquaculture, for which a monetary value can be estimated.

On the other hand, if shrimp ponds are constructed in 'reclaimed' mangrove forest, mangroves can be regarded as a direct alternative to the lucrative cultivation of shrimps. They are therefore assigned a *negative* value, the 'opportunity cost' equivalent to the economic benefit that would accrue from conversion to shrimp ponds, which is being foregone. This 'looking glass' approach takes no account of the consequences of attempting to realise this value by total removal of mangroves.

Timber, shrimps and fish are at least tangible products. What of less concrete benefits, such as protection against coastal erosion or incursion by the sea? Values can be assigned, based either on measured erosion after mangrove loss, or on the value of the protected resource such as paddy fields. Should mangroves be credited with the entire value of a commercial port, or of rice production in a coastal plain, because these might be lost if mangroves were eliminated? An alternative approach is based on assessment of the costs of the substitute, such as artificial coastal defences, which would be needed if a given service was no longer provided by mangroves.

Depending on the circumstances, the value placed on coastal protection by mangroves can be trivial, or astronomical. At one extreme, the loss of the odd hectare may not matter: at the other, loss of protective mangroves could either render a port inoperable, or require massive investment in breakwater construction. An example of this latter situation is discussed on p. 193.

The elements so far discussed are referred to as direct or indirect use values: features of mangroves whose monetary value derives from their use. Direct use values are often relatively straightforward to evaluate, indirect use values sometimes elusive. Even more intangible are non-use values. Examples of non-use values are biodiversity conservation, the ability of mangrove forests to sequester carbon or the opportunities they afford for research and education. These are generally ignored by environmental economists, but methods have been devised of assigning values to them (although it is not always clear to a mere biologist what these values actually signify).

Some valuations are shown in Table 8.3. The range of figures does reflect genuine variation in the importance of mangrove goods and services but also, to a considerable extent, differences in deciding what should be assessed, and how it should be assessed. There is no general agreement on a consistent approach to ecosystem valuation, nor any prospect of precision in measuring many of the variables. Nevertheless, assigning notional monetary values to ecosystem goods and services can be of assistance in clarifying and guiding a decision between two management alternatives within a particular mangrove system, such as whether to develop aquaculture at the expense of preservation. It is of less value in comparing two different ecosystems, where different analytical approaches are employed.

Economic valuation does, however, serve to highlight the importance of mangroves in ways which are generally not appreciated by non-biologists. A recent analysis of the major ecosystems of the world is an example of this (Costanza *et al.* 1997). The value of mangroves and tidal marshes, world-wide, was put at just under \$10 000 ha^{-1} year^{-1} (Table 8.4). The global total of \$1 648 000 000 compares with an estimated annual value of the Earth's combined ecosystems of

Table 8.3 Examples of estimates of monetary value, in US\$ ha^{-1} year^{-1}, of various mangrove systems, data from Dixon (1989), Janssen and Padilla (1996) and Ruitenbeek (1994)

	Thailand 1982	Fiji 1976	Indonesia 1992	Philippines 1996	Trinidad 1974
Forestry	30	6	67	151	70
Fisheries	130	100	117	60	125
Agriculture	165	52			
Aquaculture	−2106			−7124	
Erosion			3		
Biodiversity			15		
Local uses	230		33		
Recreation					200
Purification		5820			

Note the range of dates: to be comparable, adjustment should be made for variation, with time, in the value of the US\$. Negative values refer to option use value (see text). Gaps indicate that value was assessed as negligible, or that no attempt was made to assess value.

Table 8.4 Estimated average value of ecosystem services for mangroves and salt marsh habitats combined, from Costanza *et al.* (1997)

Goods/services	Example	Average value (US$ ha^{-1} year^{-1})
Disturbance regulation	protection against coastal erosion, typhoons	1839
Waste treatment	assimilation of effluent from shrimp ponds	6696
Habitat/refuge	nursery area for shrimps, habitat for migrating birds	169
Food production	energy source for local fisheries	466
Raw materials	timber, charcoal	162
Recreation	ecotourism	658
Total		9990

No useful data are available on the economic value of other possible ecosystem services of mangroves, such as nutrient cycling, climate regulation, genetic resources or the provision of pollinators, although some at least of these are known to be important (see, for example, p. 000).

$33 268 000 000. The true economic value of mangroves and equivalent temperate ecosystems is enormous.

Have mangroves a future?

Almost everywhere, mangroves are in retreat, in the face of the relentless pressures of deliberate destruction, over-exploitation, pollution and climatic change. If the attrition continues, in many parts of the world mangroves may be reduced to relic patches, too small to support the diversity of organisms characteristic of a thriving mangal. Is there any long-term future for mangrove ecosystems?

Very few mangrove areas are untouched by human activities. The future of mangroves must depend not on remaining isolated from human influences, but on properly regulated interactions with human demands: on effective management. For this to happen, the first step must be education. Unless planners and politicians—as well as local people—appreciate the many contributions of mangroves to human well being, and the true costs of allowing their destruction, the decline will continue. When mangroves are cleared for timber or for development, the financial gains are rapid but the losses long term. Planning must evaluate all of the costs, as well as the benefits, and look to distant horizons. And those who gain, of course, tend to have more influence on planning decisions than the local people who may lose a subsistence economy.

Some of the pressures on mangroves, such as major water diversion schemes, or sea level rise, are not amenable to local management. Here mitigation, rather than prevention, is a realistic objective. Schemes such as the replanting of mangroves in the Indus delta demonstrate how much can be achieved, at relatively low cost. Reproduction and the early settlement stages of mangroves are the most affected by environmental pressures. If these vulnerable stages can be nurtured by artificial methods, a population of adult trees may be able to survive in an environment that is no longer ideal.

With intelligent and informed management, it is possible to maintain mangroves, not just as a curiosity but as a valuable natural resource, managed sustainably so as to continue exploitation without destroying a rich and diverse habitat.

Further reading

Perhaps the best introductory book for the general reader, with some superb photographs, is

Stafford-Deitsch, J. (1996). *Mangrove. The forgotten habitat*. Immel, London.

There are few books that deal thoroughly with mangroves. The following are the most comprehensive:

Hutchings, P. and Saenger, P. (1987). *Ecology of mangroves*. University of Queensland Press, St Lucia.

Tomlinson, P.B. (1986). *The botany of mangroves*. Cambridge University Press, Cambridge.

Hutchings and Saenger is specifically about the mangroves of Australia, and Tomlinson, as the title suggests, is a thorough account of mangrove trees, with relatively little on other organisms.

Mangroves are discussed in the context of other marine ecosystems in a number of books, of which the following are excellent examples:

Barnes, R.S.K. and Mann, K.H. (ed.) (1991). *Fundamentals of aquatic ecology*. Blackwell Scientific Publications, Oxford.

Little, C. (1999). *Biology of soft shores and estuaries*. Oxford University Press, Oxford.

Bibliography

Mangroves on the Internet

The following websites contain much information on mangroves, and links to related sites:

Mangrove Home Site	http://possum.murdoch.edu.au/~mangrove/
Mangrove Action Project	http://www.earthisland.org/ei/map/map.html
Tropical Coastal Management Sites	http://www.ncl.ac.uk/tcmweb/tcm/tcmsites.htm
World Conservation Monitoring Centre	http://www.wcmc.org.uk/
Global Mangrove Protection	http://www.agri-aqua.ait.ac.th/Mangroves/

There is also a useful Mangrove Discussion Group, with a home page at

http://possum.murdoch.edu.au/~mangrove/submang.htm

References

Alongi, D.M. (1987*a*). Intertidal zonation and seasonality of meiobenthos. *Marine Biology*, **95**, 447–58.

Alongi, D.M. (1987*b*). The influence of mangrove-derived tannins on intertidal meiobenthos in tropical estuaries. *Oecologia*, **71**, 537–40.

Alongi, D.M. (1990). The ecology of tropical soft bottom benthic ecosystems. *Oceanography and Marine Biology: Annual Review*, **28**, 381–496.

Alongi, D.M. (1996). The dynamics of benthic nutrient pools and fluxes in tropical mangrove forests. *Journal of Marine Research*, **54**, 123–48.

Alongi, D.M., Boto, K.G., and Robertson, A.I. (1992). Nitrogen and phosphorus cycles. In *Tropical mangrove ecosystems* (ed. A.I. Robertson and D.M. Alongi), pp. 251–92. American Geophysical Union, Washington D.C.

Anderson, C. and Lee, S.Y. (1995). Defoliation of the mangrove *Avicennia marina* in Hong Kong: cause and consequences. *Biotropica*, **27**, 218–26.

Baird, I. and Quarto, A. (1994). The environmental and social costs of developing coastal shrimp aquaculture in Asia. In *Trade and environment in Asia-Pacific, prospects for regional cooperation*. Nautilos Institute, Berkeley.

Ball, M.C. (1988*a*). Ecophysiology of mangroves. *Trees*, **2**, 129–142.

Ball, M.C. (1988*b*). Salinity tolerance in the mangroves *Aegiceras corniculatum* and *Avicennia marina*. I. Water use in relation to growth, carbon partitioning, and salt balance. *Australian Journal of Plant Physiology*, **15**, 447–64.

Ball, M.C. and Munns, R. (1992). Plant responses to salinity under elevated atmospheric concentrations of CO_2. *Australian Journal of Botany*, **40**, 515–25.

Ball, M.C. and Pidsley, S.M. (1995). Growth responses to salinity in relation to distribution of two mangrove species, *Sonneratia alba* and *S. lanceolata*, in northern Australia. *Functional Ecology*, **9**, 77–85.

Ball, M.C., Cowan, I.R., and Farquhar, G.D. (1988). Maintenance of leaf temperature and the optimisation of carbon gain in relation to water loss in a tropical mangrove forest. *Australian Journal of Plant Physiology*, **15**, 263–76.

Banus, M.D. and Kolehmainen, S.E. (1975). Floating, rooting and growth of red mangrove (*Rhizophora mangle* L.) seedlings: effect on expansion of mangroves in south-western Puerto Rico. In *Proceedings of the international symposium on biology and management of mangroves*, (ed. G.E. Walsh, S.C. Snedaker and H.J. Teas), pp. 370–384. Institute of Food and Agricultural Sciences, University of Florida, Gainesville.

Barlow, B.A. (1966). A revision of the Loranthaceae of Australia and New Zealand. *Australian Journal of Botany*, **14**, 421–499.

Barnes, R.S.K. (1976). The osmotic behaviour of a number of grapsoid crabs with respect to their differential penetration of an estuarine system. *Journal of Experimental Biology*, **47**, 535–51.

Benner, R., Hodson, R.E., and Kirchman, D. (1988). Bacterial abundance and production on mangrove leaves during initial stages of leaching and biodegradation. *Archiv für Hydrobiologie*, **31**, 19–26.

Bertness, M.D. (1985). Fiddler crab regulation of *Spartina alterniflora* production on a New England salt marsh. *Ecology*, **66**, 1042–55.

Bhosale, L.J. and Shinde, L.S. (1983). Significance of cryptovivipary in *Aegiceras corniculatum* (L.) Blanco. In *Tasks for vegetation science. 8* (ed. H.J. Teas), pp. 123–129. W. Junk, The Hague.

Bingham, B.L. and Young, C.M. (1995). Stochastic events and dynamics of a mangrove root epifaunal community. *Marine Ecology*, **16**, 145–63.

Blasco, F., Saenger, P., and Janodet, E. (1996). Mangroves as indicators of coastal change. *Catena*, **27**, 167–78.

Boto, K.G. (1982). Nutrient and organic fluxes in mangroves. In *Mangrove ecosystems in Australia. Proceedings of the Australian National Mangrove Workshop*, (ed. B.F. Clough), pp. 239–257. Australian Institute of Marine Science, Queensland.

Boto, K.G. (1984). Waterlogged saline soils. In *The mangrove ecosystem: research methods* (ed. S.C. Snedaker and J.G. Snedaker), pp. 114–130. UNESCO, Paris.

Boto, K.G. and Wellington, J.T. (1983). Phosphorus and nitrogen nutritional status of a northern Australian mangrove forest. *Marine Ecology Progress Series*, **11**, 63–69.

Boto, K.G. and Wellington, J.T. (1984). Soil characteristics and nutrient status in a northern Australian mangrove. *Estuaries*, **7**, 61–9.

Boto, K.G., Robertson, A.I., and Alongi, D.M. (1991). Mangrove and near-shore connections—a status report from the Australian perspective. In *Proceedings of the regional symposium on living resources in coastal areas* (ed. A. Alcala), pp. 459–67. University of Philippines, Manila.

Brown, S. and Lugo, A.E. (1982). A comparison of structural and functional characteristics of saltwater and freshwater forested wetlands. In *Wetlands: Ecology and Management* (ed. B. Gopal, R.E. Turner, R.G. Wetzel, and D.F. Whigham), pp. 111–30. National Institute of Ecology and International Science Publishers, Jaipur, India.

Bruce, A.J. (1993). The occurrence of the semi-terrestrial shrimps *Merguia oligodon* (De Man 1888) and *M. rhizophorae* (Rathbun 1900) (Crustacea Decapoda Hippolytidae) in Africa. *Tropical Zoology*, **6**, 179–87.

Buck, J. and Buck, E. (1976). Synchronous fireflies. *Scientific American*, **234/5**, 74–85.

Bunt, J.S. (1995). Continental scale patterns in mangrove litter fall. *Hydrobiologia*, **295**, 135–40.

Bunt, J.S., Boto, K.G., and Boto, G. (1979). A survey method for estimating potential levels of mangrove forest primary production. *Marine Biology*, **52**, 123–8.

Burggren, W.W. and McMahon, B.R. (ed.) (1988): *Biology of the land crabs*. Cambridge University Press.

Burns, K.A., Garrity, S.D., and Levings, S.C. (1993). How many years until mangrove ecosystems recover from catastrophic oil spills? *Marine Pollution Bulletin*, **26**, 239–48.

Camilleri, J.C. (1989). Leaf choice by crustaceans in a mangrove forest in Queensland. *Marine Biology*, **102**, 453–9.

Cannicci, S., Ritosa, S., Ruwa, R.K., and Vannini, M. (1996*a*). Tree fidelity and hole fidelity in the tree crab *Sesarma leptosoma* (Decapoda, Grapsidae). *Journal of Experimental Marine Biology and Ecology*, **196**, 299–311.

Cannicci, S., Ruwa, R.K., Ritossa, S., and Vannini, M. (1996*b*). Branch fidelity in the tree crab *Sesarma leptosoma* (Decapoda, Grapsidae). *Journal of Zoology*, **238**, 795–801.

Chapman, V.J. (1976). *Mangrove vegetation*. J. Cramer, Vaduz.

Chen, R. and Twilley, R.R. (1998). A gap dynamic model of mangrove forest development along gradients of soil salinity and nutrient resources. *Journal of Ecology*, **86**, 37–51.

Chong, V.C. (1995). *The prawn-mangrove connection—fact or fallacy?* Proceedings of seminar on sustainable utilization of coastal ecosystem for agriculture, forestry and fisheries in developing regions, pp. 3–20. Malaysian Fisheries Department, Kuala Lumpur.

Chong, V.C., Sasekumar, A., Leh, M.U.C., and D'Cruz, R. (1990). The fish and prawn communities of a Malaysian coastal mangrove system, with comparisons to adjacent mud flats and inshore waters. *Estuarine, Coastal and Shelf Science*, **31**, 703–22.

Chong, V.C., Sasekumar, A., and Wolanski, E. (1996). The role of mangroves in retaining penaeid prawn larvae in Klang Strait, Malaysia. *Mangroves and Salt Marshes*, **1**, 11–22.

Clarke, P.J. (1992). Predispersal mortality and fecundity in the grey mangrove (*Avicennia marina*) in southeastern Australia. *Australian Journal of Ecology*, **17**, 161–8.

Clarke, P.J. (1993). Dispersal of grey mangrove (*Avicennia marina*) propagules in southeastern Australia. *Aquatic Botany*, **45**, 195–204.

Clarke, P.J. and Myerscough, P.J. (1993). The intertidal distribution of the gray mangrove (*Avicennia marina*) in southeastern Australia—the effects of physical conditions, interspecific competition, and predation on propagule establishment and survival. *Australian Journal of Ecology*, **18**, 307–15.

Clay, R.E., and Andersen, A.N. (1996). Ant fauna of a mangrove community in the Australian seasonal tropics, with particular reference to zonation. *Australian Journal of Zoology*, **44**, 521–33.

Clayton, D.A. (1993). Mudskippers. *Oceanography and Marine Biology Annual Review*, **31**, 507–77.

Clough, B.F. (1992). Primary productivity and growth of mangrove forests. In *Tropical mangrove ecosystems* (ed. A.I. Robertson and D.M. Alongi), pp. 225–49. American Geophysical Union, Washington D.C.

Clough, B.F. and Scott, K. (1989). Allometric relationships for estimating above-ground biomass in six mangrove species. *Forest Ecology and Management*, **27**, 117–27.

Clough, B.F., Boto, K.G., and Attiwill, P.M. (1983). Mangroves and sewage: a re-evaluation. In *Tasks for vegetation science, 8* (ed. H.J. Teas), pp. 151–61. W. Junk, The Hague.

Cook, L.M. (1986). Site selection in a polymorphic mangrove snail. *Biological Journal of the Linnean Society*, **29**, 101–13.

Cook, L.M. and Freeman, P.M. (1986). Heating properties of morphs of the mangrove snail, *Littoraria pallescens*. *Biological Journal of the Linnean Society*, **29**, 295–300.

Cook, L.M., Currey, J.D., and Sarsam, V.H. (1985). Differences in morphology in relation to microhabitat in littorinid species from a mangrove in Papua New Guinea. *Journal of Zoology*, **206**, 297–310.

Costanza, R., d'Arge, R., de Groot, R., Farber, S., Grasso, M., Hannon, B., *et al.* (1997). The value of the world's ecosystem services and natural capital. *Nature*, **387**, 253–60.

Crane, J. (1975). *Fiddler crabs of the world, Ocypodidae: genus Uca*. Princeton University Press, Princeton.

Curran, M. (1985). Gas movements in the roots of *Avicennia marina* (Forsk.) Vierh. *Australian Journal of Plant Physiology*, **12**, 97–108.

Dame, R.F. and Allen, D.M. (1996). Between estuaries and the sea. *Journal of Experimental Marine Biology and Ecology*, **200**, 169–185.

Davie, P.J.F. (1982). A preliminary checklist of Brachyura (Crustacea: Decapoda) associated with Australian mangrove forests. *Operculum*, **5**, 204–7.

Dierenfeld, E.S., Koontz, F.W., and Goldstein, R.S. (1992). Feed intake, digestion and passage of the Proboscis monkey (*Nasalis larvatus*) in captivity. *Primates*, **33**, 399–405.

Dittel, A.I., Epifanio, C.E., and Lizano, O. (1991). Flux of larvae in a mangrove creek in the Gulf of Nicoya, Costa Rica. *Estuarine, Coastal and Shelf Science*, **32**, 129–40.

Dixon, J.A. (1989). Valuation of mangroves. *Tropical Coastal Area Management*, **4**, 1–6.

Duarte, C.M. and Cebrián, J. (1996). The fate of marine autotrophic production. *Limnology and Oceanography*, **41**, 1758–66.

Duke, N.C. (1992). Mangrove floristics and biogeography. In *Tropical mangrove ecosystems* (ed. A.I. Robertson and D.M. Alongi), pp. 63–100. American Geophysical Union, Washington D.C.

Duke, N.C. (1995). Genetic diversity, distributional barriers and rafting continents—more thoughts on the evolution of mangroves. *Hydrobiologia*, **295**, 167–81.

Duke, N.C., Pinzón, Z.S., and Prada, M.C. (1997). Large-scale damage to mangrove forests following two large oil spills in Panama. *Biotropica*, **29**, 2–14.

Dunson, W.A. (1970). Some aspects of electrolyte and water balance in three estuarine reptiles, the diamond back terrapin, American and 'salt water' crocodiles. *Comparative Biochemistry and Physiology*, **32**, 161–74.

Dunson, W.A. (1974). Salt gland secretion in a mangrove monitor lizard. *Comparative Biochemistry and Physiology*, **47A**, 1245–55.

Dunson, W.A. (1978). Role of the skin in sodium and water exchange of aquatic snakes placed in sea water. *American Journal of Physiology*, **235**, R151–R9.

Dye, A.H. and Lasiak, T.A. (1986). Microbenthos, meiobenthos and fiddler crabs: trophic interactions in a tropical mangrove sediment. *Marine Ecology Progress Series*, **32**, 259–64.

Dye, A.H. and Lasiak, T.A. (1987). Assimilation efficiencies of fiddler crabs and deposit-feeding gastropods from tropical mangrove sediments. *Comparative Biochemistry and Physiology*, **87A**, 341–4.

Edney, E.B. (1961). The water and heat relationships of fiddler crabs (*Uca* spp.). *Transactions of the Royal Society of South Africa*, **36**, 71–91.

Elliott, A.B. and Karunakaran, L. (1974). Diet of *Rana cancrivora* in fresh water and brackish water environments. *Journal of Zoology*, **174**, 203–15.

Ellison, A.M. and Farnsworth, E.J. (1996). Anthropogenic disturbance of Caribbean mangrove ecosystems: past impacts, present trends, and future predictions. *Biotropica*, **28(4a)**, 549–65.

Ellison, A.M., Farnsworth, E.J., and Twilley, R.R. (1996). Facultative mutualism between red mangroves and root-fouling sponges in Belizean mangal. *Ecology*, **77**, 2431–44.

Ellison, J.C. and Stoddart, D.R. (1991). Mangrove ecosystem collapse during predicted sea-level rise: Holocene analogues and implications. *Journal of Coastal Research*, **7**,151–65.

Elmqvist, T. and Cox, P.A. (1996). The evolution of vivipary in flowering plants. *Oikos*, **77**, 3–9.

Emmerson, W.D. and McGwynne, M.E. (1992). Feeding and assimilation of mangrove leaves by the crab *Sesarma meinerti* de Man in relation to leaf litter production in Mgazama, a warm-temperate southern African mangrove swamp. *Journal of Experimental Marine Biology and Ecology*, **157**, 41–53.

English, S., Wilkinson, L., and Baker, V. (eds) (1997). *Survey manual for tropical marine resources*, 2nd edn. Australian Institute of Marine Science, Townsville, Queensland.

Epifanio, C. (1988). Transport of crab larvae between estuaries and the continental shelf. In *Lecture notes on coastal and estuarine systems, 22*, (ed. B.-O. Jansson), pp. 291–305. Springer-Verlag, Berlin.

Eshky, A.A., Atkinson, R.J.A., and Taylor, A.C. (1995). Physiological ecology of crabs from Saudi Arabian mangrove. *Marine Ecology Progress Series*, **126**, 83–95.

Ewel, K.C., Twilley, R.R., and Ong, J.E. (1998). Different kinds of mangrove forests provide different goods and services. *Global Ecology and Biogeography Letters*, **7**, 83–94.

FAO (1994). *Mangrove forest management guidelines*. Food and Agricultural Organization, Rome.

Farah, A. and Meynell, P.J. (1992). Sea level rise—possible impacts on the Indus Delta, Pakistan. *Korangi ecosystem project: issues paper 2*, pp. 1–20. International Union for the Conservation of Nature, Pakistan.

Farnsworth, E.J. and Ellison, A.M. (1991). Patterns of herbivory in Belizean mangrove swamps. *Biotropica*, **13/4B**, 555–67.

Farnsworth, E.J. and Ellison, A.M. (1996). Scale-dependent spatial and temporal variability in biogeography of mangrove root epibiont communities. *Ecological Monographs*, **66**, 45–66.

Farnsworth, E.J. and Farrant, J.M. (1998). Reductions in abcsisic acid are linked with viviparous reproduction in mangroves. *American Journal of Botany*, **85**, 760–9.

Felgenhauer, B.E. and Abele, L.G. (1983). Branchial water movement in the grapsid crab *Sesarma reticulatum* (Say). *Journal of Crustacean Biology*, **3**, 187–95.

Feller, I.C. (1995). Effects of nutrient enrichment on growth and herbivory of dwarf red mangrove (*Rhizophora mangle*). *Ecological Monograph*, **65**, 477–705.

Feller, I.C. (1996). Effects of nutrient enhancement on leaf anatomy of dwarf *Rhizophora mangle* L. (red mangrove). *Biotropica*, **28**, 13–22.

Fenchel, T., Esteban, G.F. and Finlay, B.J. (1997). Local versus global diversity of microorganisms: cryptic diversity of ciliated protozoa. *Oikos*, **80**, 220–5.

Field, C.D. (1984). Movement of ions and water into the xylem sap of tropical mangroves. In *Physiology and management of mangroves* (ed. H.J. Teas), pp. 49–52. Junk, The Hague.

Field, C.D. (1995). Impact of expected climate change on mangroves. *Hydrobiologia*, **295**, 75–81.

Fleming, M., Lin, G., and Sternberg, L.da S.L. (1990). Influence of mangrove detritus in an estuarine ecosystem. *Bulletin of Marine Science*, **47**, 663–9.

Ford, J. (1982). Origin, evolution and speciation of birds specialized to mangroves in Australia. *Emu*, **82**, 12–23.

Franklin, C.E. and Grigg, G.C. (1993). Increased vascularity of the lingual salt-glands of the estuarine crocodile, *Crocodylus porosus*, kept in hyperosmotic salinity. *Journal of Morphology*, **218**, 143–51.

Furukawa, K. and Wolanski, E. (1996). Sedimentation in mangrove forests. *Mangroves and Salt Marshes*, **1**, 3–10.

Furukawa, K., Wolanski, E., and Mueller, H. (1997). Currents and sediment transport in mangrove forests. *Estuarine, Coastal and Shelf Science*, **44**, 301–10.

Gan, B.K. (1993). Forest management in Matang. In *Living coastal resources. Proceedings of workshop on mangrove fisheries and connections* (ed. A. Sasekumar), pp. 15–26. Australian International Assistance Bureau (AIDAB).

Gee, J.M. and Somerfield, P.J. (1997). Do mangrove diversity and leaf litter decay promote meiofaunal diversity? *Journal of Experimental Marine Biology and Ecology*, **218**, 13–33.

Giddins, R.L., Lucas, J.S., Neilson, M.J. and Richards, G.N. (1986). Feeding ecology of the mangrove crab *Neosarmatium smithi* (Crustacea: Decapoda: Sesarmidae). *Marine Ecology Progress Series*, **33**, 147–55.

Gill, A.M. and Tomlinson, P.B. (1969). Studies on the growth of red mangrove (*Rhizophora mangle* L.). 1. Habit and general morphology. *Biotropica*, **1**, 1–9.

Gill, A.M. and Tomlinson, P.B. (1975). Aerial roots: an array of forms and functions. In *The development and function of roots* (ed. J.G. Torrey and D.T. Clarkson), pp. 237–60. Academic Press, London.

Gill, A.M. and Tomlinson, P.B. (1977). Studies on the growth of red mangrove (*Rhizophora mangle* L.). 4. The adult root system. *Biotropica*, **9**, 145–55.

Gomez, M.A. and Winkler, S. (1991). Bromeliads from mangroves on the Pacific coast of Guatemala. *Revista de Biologia Tropical*, **39**, 207–14.

Gong, W.K. and Ong, J.E. (1990). Plant biomass and nutrient flux in a managed mangrove forest in Malaysia. *Estuarine, Coastal and Shelf Science*, **31**, 519–30.

Gong, W.K. and Ong, J.E. (1995). The use of demographic studies in mangrove silviculture. *Hydrobiologia*, **295**, 255–61.

González-Farias, F. and Mee, L.D. (1988). Effect of mangrove humic-like substances on biodegradation rate of detritus. *Journal of Experimental Marine Biology and Ecology*, **119**, 1–13.

Gopinath, N. and Gabriel, P. (1997). Management of living resources in the Matang Mangrove Reserve, Perak, Malaysia. *Intercoast Network*, March, 23–4.

Gordon, M.S. and Tucker, V.A. (1965). Osmotic regulation in the tadpoles of the crab-eating frog (*Rana cancrivora*). *Journal of Experimental Biology*, **42**, 437–45.

Gordon, M.S., Schmidt-Nielsen, K., and Kelly, H.M. (1961). Osmotic regulation in the crab-eating frog (*Rana cancrivora*). *Journal of Experimental Biology*, **38**, 659–78.

Gordon, M.S., Ng, W.W.S. and Yip, A.Y.W. (1978). Aspects of the physiology of terrestrial life in amphibious fishes. III. The Chinese mudskipper *Periophthalmus cantonensis*. *Journal of Experimental Biology*, **72**, 57–75.

Graves, J.D. and Reavey, D. (1996). *Global environmental change: plants, animals and communities*. Longman, Harlow.

Gray, I.E. (1957). A comparative study of the gill area of crabs. *Biological Bulletin*, **112**, 34–42.

Gunasekar, M (1993). Changes in chlorophyll content, photosynthetic rate and saccharides level in developing *Rhizophora* hypocotyls. *Photosynthetica*, **29**, 635–8.

Hamilton, L.S. and Murphy, D.H. (1988). Use and management of Nipa Palm (*Nypa fruticans*, Arecaceae): a review. *Economic Botany*, **42**, 206–13.

Hemminga, M.A., Slim, F.J., Kazungu, J., Ganssen, G.M., Nieuwenhuize, J., and Kruyt, N.M. (1994). Carbon outwelling from a mangrove forest with adjacent seagrass beds and coral reefs (Gazi Bay, Kenya). *Marine Ecology Progress Series*, **106**, 291–304.

Higgins, R.P. and Thiel, H. (eds) (1988). *Introduction to the study of meiofauna*. Smithsonian Institution.

Hogarth, P.J. (1999) The decline in Indus Delta mangroves: causes, consequences, and cures. *Proceedings of International Conference on Mangroves, Kuwait, 1998.* Kuwait Institute for Scientific Research, Kuwait.

Houbrick, R.S. (1991). Systematic review and functional morphology of the mangrove snails *Terebralia* and *Telescopium* (Potamididae, Prosobranchia). *Malacologia*, **33**, 289–338.

Hutchings, P. and Saenger, P. (1987). *Ecology of mangroves.* University of Queensland Press, St Lucia.

Huxley, C.R. (1978). The ant-plants *Myrmecodia* and *Hydnophytum* (Rubiaceae) and the relationships between their morphology, ant occupants, physiology and ecology. *New Phytologist*, **80**, 231–68.

Hyde, K.D. and Lee, S.Y. (1995). Ecology of mangrove fungi and their role in nutrient cycling: what gaps occur in our knowledge? *Hydrobiologia*, **295**, 107–18.

Icely, J.D. and Jones, D.A. (1978). Factors affecting the distribution of the genus *Uca* (Crustacea: Ocypodidae) on an East African shore. *Estuarine and Coastal Marine Science*, **6**, 315–25.

Ishimatsu, A., Khoo, K.H., and Takita, T. (1998). Deposition of air in burrows of tropical mudskippers as an adaptation to the hypoxic mudflat environment. *Science Progress*, **81**, 289–97.

Jackson, K., Butler, D.G., and Brooks, D.R. (1996). Habitat and phylogeny influence salinity discrimination on crocodilians—implications for osmoregulatory physiology and historical biogeography. *Biological Journal of the Linnean Society*, **58**, 371–83.

Janssen, R. and Padilla, J.E. (1996). Valuation and evaluation of management alternatives for the Pagbilao mangrove forest. *CREED Working Paper*, **9**, 1–47. International Institute for Environment and Development, Amsterdam.

Janzen, D.H. (1974). Epiphytic myrmecophytes in Sarawak: mutualism through the feeding of plants by ants. *Biotropica*, **6**, 237–59.

Janzen, D.H. (1985). Mangroves: where's the understory? *Journal of Tropical Ecology*, **1**, 89–92.

Jones, C.G., Lawton, J.H., and Shachak, M. (1994). Organisms as ecosystem engineers. *Oikos*, **69**, 373–86.

Jones, D.A. (1984). Crabs of the mangal ecosystem. In *Hydrobiology of the mangal.* (ed. F.D. Por and I. Dor), pp. 89–109. W. Junk, The Hague.

Joshi, G.V., Pimplaskar, M., and Bhosale, L.J. (1972). Physiological studies in germination of mangroves. *Botanica marina*, **15**, 91–95.

Katz, L.C. (1980). Effects of burrowing by the fiddler crab, *Uca pugnax* (Smith). *Estuarine and Coastal Marine Science*, **11**, 233–7.

Kautsky, N., Berg, H., Folke, C., Larsson, J., and Troell, M. (1997). Ecological footprint for assessment of resource use and development limitations in shrimp and tilapia aquaculture. *Aquaculture Research*, **28**, 753–66.

Klekowski, E.J., Lowenfeld, R., and Hepler, P.K. (1994). Mangrove genetics. II. Outcrossing and lower spontaneous mutation rates in Puerto Rican *Rhizophora. International Journal of Plant Science*, **155**, 373–81.

Kwok, P.W. and Lee, S.Y. (1995). The growth performances of two mangrove crabs, *Chiromanthes bidens* and *Parasesarma plicata* under different leaf litter diets. *Hydrobiologia*, **295**, 141–8.

Lee, S.Y. (1991). Herbivory as an ecological process in a *Kandelia candel* (Rhizophoraceae) mangal in Hong Kong. *Journal of Tropical Ecology*, **7**, 337–48.

Lee, S.Y. (1997). Potential trophic importance of the faecal material of the mangrove sesarmine crab *Sesarma messa. Marine Ecology Progress Series*, **159**, 275–84.

Lefebvre, G. and Poulin, B. (1997). Bird communities in Panamanian black mangroves: potential effects of physical and biotic factors. *Journal of Tropical Ecology*, **13**, 97–113.

Leh, C.M.U. and Sasekumar, A. (1985). The food of sesarmid crabs in Malaysian mangrove forests. *Malayan Nature Journal*, **39**, 135–45.

Lewis, R.R. (1983). Impact of oil spills on mangrove forests. In *Tasks for vegetation science, 8* (ed. H.J. Teas), pp. 171–83. W. Junk, The Hague.

Lin, G.H. and Sternberg, L.D.L. (1994). Utilization of surface water by Red Mangrove (*Rhizophora mangle* L.)—an isotopic study. *Bulletin of Marine Science*, **54**, 94–102.

Lin, G.H., Banks, T., and Sternberg, L. da S.L.. (1991). Variations in $\delta^{13}C$ values for the seagrass *Thalassia testudinum* and its relations to mangrove carbon. *Aquatic Botany*, **40**, 333–341.

Lötschert, W. and Liemann, F. (1967). Die Salzspeicherung im Keimling von *Rhizophora mangle* L. während der Entwicklung auf der Mutterpflanze. *Planta*, **77**, 142–156.

Lugo, A.E. (1980). Mangrove ecosystems: successional or steady state? *Biotropica*, **12s**, 65–72.

Lugo, A.E. and Snedaker, S.C. (1974). The ecology of mangroves. *Annual Reviews of Ecology and Systematics*, **5**, 39–64.

Lugo, A.E., Sell, M., and Snedaker, S.C. (1976). Mangrove ecosystem analysis. In *Systems analysis and simulation in ecology* (ed. C. Patten), pp. 113–45. Academic Press, New York.

Macintosh, D.J. (1982). Ecological comparisons of mangrove swamp and salt marsh fiddler crabs. In *Wetlands, ecology and management* (ed. B. Gopal, R.E. Turner, R.G. Wetzer and D.F. Waigham), pp. 243–57. International Science Publications, Jaipur.

Macintosh, D.J. (1984). Ecology and productivity of Malaysian mangrove crab popula- tions. In *Proceedings of the Asian symposium on the mangrove environment—research and manage- ment, 1984*, pp. 354–77.

Macintosh, D.J. (1988). The ecology and physiology of decapods of mangrove swamps. *Symposium of the Zoological Society of London*, **59**, 315–41.

Macintosh, D.J. (1996). Mangroves and coastal aquaculture: doing something positive for the environment. *Aquaculture Asia*, October–December, 3–8.

Macnae, W. (1968). A general account of the fauna and flora of the mangrove swamps and forests in the Indo-Pacific Region. *Advances in Marine Biology*, **6**, 73–270.

Mann, K.H. (1982). *Ecology of coastal waters. A systems approach*. Blackwell, London.

Marguillier, S., van der Velde, G., Dehairs, F., Hemminga, M.A., and Rajagopal, S. (1997). Trophic relationships in an interlinked mangrove-seagrass ecosystem as traced by $\delta^{13}C$ and $\delta^{15}N$. *Marine Ecology Progress Series*, **151**, 115–21.

Mazzotti, F.J. and Dunson, W.A. (1984). Adaptations of *Crocodylus acutus* and *Alligator* for life in saline water. *Comparative Biochemistry and Physiology*, **79A**, 641–6.

McCoy, E.D., Mushinsky, H.R., Johnson, D., and Meshaka, W.E. (1996). Mangrove damage caused by Hurricane Andrew on the southwestern coast of Florida. *Bulletin of Marine Science*, **59**, 1–8.

McGrady-Steed, J., Harris, P.M., and Morin, P.J. (1997). Biodiversity regulates ecosystem predictability. *Nature*, **390**, 162–5.

McGuinness, K.A. (1997). Dispersal, establishment and survival of *Ceriops tagal* propa- gules in a north Australian mangrove forest. *Oecologia*, **109**, 80–7.

McIvor, C.C. and Smith, T.J. (1996). Differences in the crab fauna of mangrove areas at southwest Florida and a northeast Australia location: implications for leaf litter processing. *Estuaries*, **18**, 591–7.

McKee, K.L. (1993). Soil physicochemical patterns and mangrove species distribution— reciprocal effects? *Journal of Ecology*, **81**, 477–87.

McKee, K.L. (1995). Mangrove species distribution and propagule predation in Belize: an exception to the dominance-predation hypothesis. *Biotropica*, **27**, 334–45.

McKee, K.L. and Mendelssohn, I.A. (1987). Root metabolism in the black mangrove (*Avicennia germinans* (L.) L): response to hypoxia. *Environmental and Experimental Botany*, **27**, 147–156.

McKenzie, N.L. and Rolfe, J.K. (1986). Structure of bat guilds in the Kimberley mangroves, Australia. *Journal of Animal Ecology*, **55**, 401–20.

McMillan, C. (1975). Adaptive differentiation to chilling in mangrove populations. In *Proceedings of the international symposium on biology and management of mangroves* (ed. G.E. Walsh, S.C. Snedaker and H.J. Teas), pp. 62–8. Institute of Food and Agricultural Sciences, University of Florida, Gainesville.

Micheli, F., Gherardi, F., and Vannini, M. (1991). Feeding and burrowing ecology of two East African mangrove crabs. *Marine Biology*, **111**, 247–54.

Miller, D.C. (1961). The feeding mechanisms of fiddler crabs, with ecological considerations of feeding adaptations. *Zoologica*, **46**, 89–100.

Milliman, J.D., Quraishee, G.S., and Beg, M.A.A. (1984). Sediment discharge from the Indus River to the Ocean: past, present and future. In *Marine geology and oceanography of Arabian Sea and coastal Pakistan*, (ed. B.U. Haq and J.D. Milliman), pp. 65–70. Van Nostrand Reinhold, New York.

Moon, G.J., Clough, B.F., Peterson, C.A., and Allaway, W.G. (1986). Apoplastic and symplastic pathways in *Avicennia marina* (Forsk.) Vierh. roots revealed by fluorescent tracer dyes. *Australian Journal of Plant Physiology*, **13**, 637–48.

Moran, M.A., Wicks, R.J., and Hodson, R.E. (1991). Export of dissolved organic matter from a mangrove swamp ecosystem—evidence from natural fluorescence, dissolved lignin phenols, and bacterial secondary production. *Marine Ecology Progress Series*, **76**, 175–84.

Naeem, S. and Li, S. (1997). Biodiversity enhances ecosystem reliability. *Nature*, **390**, 507–9.

Nielsen, M.G. (1997). Nesting biology of the mangrove mud-nesting ant *Polyrhachis sokolova* Forel (Hymenoptera, Formicidae) in northern Australia. *Insectes Sociaux*, **44**, 15–21.

Neilson, M.J., Giddins, R.L., and Richards, G.N. (1986). Effects of tannins on the palatability of mangrove leaves to the tropical sesarmid crab *Neosarmatium smithi*. *Marine Ecology Progress Series*, **34**, 185–6.

Noske, R.A. (1995). The ecology of mangrove forest birds in Peninsular Malaysia. *Ibis*, **137**, 250–63.

Noske, R.A. (1996). Abundance, zonation and foraging ecology of birds in mangroves of Darwin Harbour, Northern Territory. *Wildlife Research*, **23**, 443–74.

Odum, W.E. and Heald, E.J. (1975). The detritus-based food web of an estuarine mangrove community. In *Estuarine research* (ed. L.E. Cronin), pp. 265–86. Academic Press, New York.

Ogden, J.C. and Gladfelter, E.H. (eds) (1983). *Coral Reefs, seagrass beds and mangroves: their interaction in the coastal zones of the Caribbean*. UNESCO, Paris.

Olafsson, E. and Ndaro, S.G.M. (1997). Impact of the mangrove crabs *Uca annulipes* and *Dotilla fenestrata* on meiobenthos. *Marine Ecology Progress Series*, **158**, 225–32.

Ong, J.E. (1993). Mangroves—a carbon source and sink. *Chemosphere*, **27**, 1097–1107.

Ong, J.E. (1995). The ecology of mangrove conservation and management. *Hydrobiologica*, **295**, 343–351.

Ong, J.E., Gong, W.K., Wong, C.H., and Dhanarajan, G. (1984). Contribution of aquatic productivity in managed mangrove ecosystem in Malaysia. In *Proceedings of the Asian symposium on mangrove environment research and management*, (ed. E. Soepadmo, A.N. Rao and D.J. Macintosh), pp. 209–15. University of Malaya, Kuala Lumpur.

Ono, Y. (1965). On the ecological distribution of ocypodid crabs in the estuary. *Memoirs of the Faculty of Science of Kyushu University*, **4**, 1–60.

Onuf, C.P., Teal, J.M., and Valiela, I. (1977). Interactions of nutrients, plant growth and herbivory in a mangrove ecosystem. *Ecology*, **58**, 514–26.

Osborne, D.J. and Berjak, P. (1997). The making of mangroves: the remarkable pioneering role played by the seed of *Avicennia marina*. *Endeavour*, **21**, 143–7.

Osborne, K. and Smith, T.J. (1990). Differential predation on mangrove propagules in open and closed canopy forest habitats. *Vegetatio*, **89**, 1–6.

Pang, S.C. (1989). *Traditional fishing activities in the mangrove ecosystems of Sarawak*. Department of Fisheries, Kuala Lumpur.

Pannier, F. and Fraíno de Pannier, R. (1975). Physiology of vivipary in *Rhizophora mangle*. In *Proceedings of the international symposium on biology and management of mangroves* (ed. G.E. Walsh, S.C. Snedaker and H.J. Teas), pp. 632–9. Institute of Food and Agricultural Sciences, University of Florida, Gainesville.

Parani, M., Lakshmi, M., Elango, S., Ram, N., Anuratha, C.S., and Parida, A. (1997). Molecular phylogeny of mangroves II. Intra- and inter-specific variation in *Avicennia* revealed by RAPD and RFLP markers. *Genome*, **40**, 487–95.

Passioura, J.B., Ball, M.C., and Knight, J.H. (1992). Mangroves may salinize the soil and in so doing limit their transpiration rate. *Functional Ecology*, **6**, 476–81.

Pernetta, J.C. (1993). *Mangrove forests, climate change and sea level rise: hydrological influences on community structure and survival, with examples from the Indo-West Pacific*. International Union for the Conservation of Nature (IUCN), Gland.

Perry, D.M. (1988). Effects of associated fauna on growth and productivity in the red mangrove. *Ecology*, **6**, 1064–75.

Peters, E.C., Gassman, N.J., Firman, J.C., Richmond, R.H., and Power, E.A. (1997). Ecotoxicology of tropical marine ecosystems. *Environmental Toxicology and Chemistry*, **16**, 12–40.

Phillips, R.C. and McRoy, C.P. (1980). *Handbook of seagrass biology: an ecosystem perspective*. Garland STPM Press, New York.

Pidcock, S., Taplin, L.E., and Grigg, G.C. (1997). Differences in renal-cloacal function between *Crocodylus porosus* and *Alligator mississippiensis* have implications for crocodilian evolution. *Journal of Comparative Physiology B*, **167**, 153–8.

Pinto, L. and Wignarajah, S. (1980). Some ecological aspects of the edible oyster *Crassostrea cucullata* (Born) occurring in association with mangroves in Negombo Lagoon, Sri Lanka. *Hydrobiologia*, **69**, 11–19.

Plaziat, J.C. (1984). Mollusk distribution in the mangal. In *Hydrobiology of the mangal* (ed. F.D. Por and I. Dor), pp. 111–43. Junk, The Hague.

Plaziat, J.C. (1995). Modern and fossil mangroves and mangals: their climatic and biogeographic variability. In *Marine palaeoenvironmental analysis from fossils* (ed. D.W.J. Bosence and P.A. Allison), pp. 73–96. Geological Society Special Publication 83, London.

Price, A.R.G., Medley, P.A.H., McDowall, R.J., Dawson-Shepherd, A.R., Hogarth, P.J., and Ormond, R.F.G. (1987). Aspects of mangal ecology along the Red Sea coast of Saudi Arabia. *Journal of Natural History*, **21**, 449–64.

Primavera, J.H. (1997a). Fish predation on mangrove-associated penaeids. The role of structures and substrate. *Journal of Experimental Marine Biology and Ecology*, **215**, 205–216.

Primavera, J.H. (1997*b*). Socio-economic impacts of shrimp culture. *Aquaculture Research*, **28**, 815–27.

Putz, F.E. and Chan, H-T. (1986). Tree growth, dynamics and productivity in a mature mangrove forest in Malaysia. *Forest Ecology and Management*, **17**, 211–30.

Qureshi, M.T. (1996). Restoration of mangroves in Pakistan. In *Restoration of mangrove ecosystems* (ed. C. Field), pp. 126–42. International Timber Organisation and International Society for Mangrove Ecosystems, Okinawa, Japan.

Rabinowitz, D. (1978*a*). Early growth of mangrove seedlings in Panama, and an hypothesis concerning the relationship of dispersal and zonation. *Journal of Biogeography*, **5**, 113–33.

Rabinowitz, D. (1978*b*). Dispersal properties of mangrove propagule. *Biotropica*, **10**, 47–57.

Rabinowitz, D. (1978*c*). Mortality and initial propagule size in mangrove seedlings in Panama. *Journal of Ecology*, **66**, 45–61.

Ricklefs, R.E. and Latham, R.E. (1993). Global patterns of diversity in mangrove floras. In *Species diversity in ecological communities* (ed. R.E. Ricklefs and D. Schluter), pp. 215–29. University of Chicago Press, Chicago.

Ricklefs, R.E. and Schluter, D. (1993). Species diversity: regional and historical influences. In *Species diversity in ecological communities* (ed. R.E. Ricklefs and D. Schluter), pp 350–63. University of Chicago Press, Chicago.

Ridd, P.V. (1996). Flow through animal burrows in mangrove creeks. *Estuarine, Coastal and Shelf Science*, **43**, 617–25.

Ridd, P.V. and Samm, R. (1996). Profiling groundwater salt concentrations in mangrove swamps and tropical salt flats. *Estuarine, Coastal and Shelf Science*, **43**, 627–35.

Rivera Monroy, V.H., Day, J.W., Twilley, R.R., and Veraherrera, F. (1995). Flux of nitrogen and sediment in a fringe mangrove forest in Terminos Lagoon, Mexico. *Estuarine, Coastal and Shelf Science*, **40**, 139–60.

Robertson, A.I. (1986). Leaf-burying crabs: their influence on energy flow and export from mixed mangrove forests (*Rhizophora* spp.) in northeastern Australia. *Journal of Experimental Marine Biology and Ecology*, **102**, 237–48.

Robertson, A.I. (1988). *Food chains in tropical Australian mangrove habitats: a review of recent research*. UNDP/UNESCO Symposium on New Perspectives in Research and Management of Mangrove Ecosystems, New Delhi.

Robertson, A.I. (1991). Plant-animal interactions and the structure and function of mangrove forest ecosystems. *Australian Journal of Ecology*, **16**, 433–43.

Robertson, A.I. and Blaber, S.J.M. (1992). Plankton, epibenthos and fish communities. In *Tropical mangrove ecosystems* (ed. A.I. Robertson and D.M. Alongi), pp. 173–224. American Geophysical Union, Washington D.C.

Robertson, A.I. and Daniel, P.A. (1989). The influence of crabs on litter processing in high intertidal mangrove forests in tropical Australia. *Oecologia*, **78**, 191–8.

Robertson, A.I. and Duke, N.C. (1987). Insect herbivory on mangrove leaves in North Queensland. *Australian Journal of Ecology*, **12**, 1–7.

Robertson, A.I. and Phillips, M.J. (1995). Mangroves as filters of shrimp pond effluent: prediction and biochemical research needs. *Hydrobiologia*, **295**, 11–321.

Robertson, A.I., Giddins, R., and Smith, T.J. (1990). Seed predation by insects in tropical mangrove forests: extent and effects on seed viability and the growth of seedlings. *Oecologia*, **83**, 213–9.

Robertson, A.I., Alongi, D.M., and Boto, K.G. (1992). Food chains and carbon fluxes. In *Tropical Mangrove Ecosystems* (ed. A.I. Robertson and D.M. Alongi), pp. 293–326. American Geophysical Union, Washington D.C.

Rodelli, M.R., Gearing, J.N., Gearing, P.J., Marshall, N., and Sasekumar, A. (1984). Stable isotope ratio as a tracer of mangrove carbon in Malaysian ecosystems. *Oecologia*, **61**, 326–33.

Rosenzweig, M.L. (1995). *Species diversity in space and time*. Cambridge University Press, Cambridge.

Ruitenbeek, H.J. (1994). Modelling economy-ecology linkages in mangroves: economic evidence for promoting conservation in Bintuni Bay, Indonesia. *Ecological Economics*, **10**, 233–47.

Saenger, P. and Bellan, M.F. (1995). *The mangrove vegetation of the Atlantic coast of Africa. A review*. Laboratoire d'Ecologie Terrestre, Toulouse.

Saenger, P. and Snedaker, S.C. (1993). Pantropical trends in mangrove above-ground biomass and annual litterfall. *Oecologia*, **96**, 293–9.

Saintilan, N. (1997). Above- and below-ground biomasses of two species of mangrove on the Hawkesbury River estuary, New South Wales. *Marine and Freshwater Research*, **48**, 147–52.

Salter, R.E., MacKenzie, N.A., Nightingale, N., Aken, K.M., and Chai, P.P.K. (1985). Habitat use, ranging behaviour, and food habits of the Proboscis monkey, *Nasalis larvatus* (van Wurmb), in Sarawak. *Primates*, **26**, 436–51.

Sasekumar, A. (1994). Meiofauna of a mangrove shore on the west coast of Peninsular Malaysia. *Raffles Bulletin of Zoology*, **42**, 901–15.

Sasekumar, A., Ong, T.L., and Thong, K.L. (1984). Predation of mangrove fauna by marine fishes. In *Proceedings of the Asian symposium on the mangrove environment–research and management*, (ed. E. Soepadmo, A.N. Rao and D.J. Macintosh), pp. 378–84. University of Malaya, Kuala Lumpur.

Sasekumar, A., Chong, C.V., Leh, M.U., and D'Cruz, R. (1992). Mangroves as a habitat for fish and prawns. *Hydrobiologia*, **247**, 195–207.

Scholander, P.F. (1955). How mangroves desalinate seawater. *Physiologia Plantarum*, **21**, 251–61.

Scholander, P.F., van Dam, L., and Scholander, S.I. (1955). Gas exchange in the roots of mangroves. *American Journal of Botany*, **42**, 92–8.

Schrijvers, J. and Vincx, M. (1997). Cage experiments in an East African mangrove forest: a synthesis. *Journal of Sea Research*, **38**, 123–33.

Schrijvers, J., Schallier, R., Silence, J., Okondo, J.P., and Vincx, M. (1997). Interactions between epibenthos and meiobenthos in a high intertidal *Avicennia marina* mangrove forest. *Mangroves and Salt Marshes*, **1**, 137–54.

Semeniuk, V. (1983). Mangrove distribution in northern Australia in relationship to regional and local freshwater seepage. *Vegetatio*, **53**, 11–31.

Simberloff, D. (1976). Experimental zoogeography of islands: effects of island size. *Ecology*, **57**, 629–48.

Simberloff, D., Brown, B.J., and Lowrie, S. (1978). Isopod and insect borers may benefit Florida mangroves. *Science*, **201**, 630–2.

Simberloff, D.S. (1983). Mangroves. In *Costa Rican natural history* (ed. D.H. Janzen), pp. 273–6. University of Chicago Press, Chicago.

Simberloff, D.S. and Wilson, E.O. (1969). Experimental zoogeography of islands: the colonization of empty islands. *Ecology*, **50**, 278–96.

Simberloff, D.S. and Wilson, E.O. (1970). Experimental zoogeography of islands: a two-year record of colonization. *Ecology*, **51**, 934–7.

Slim, F.J., Hemminga, M.A., Ochieng, C., Jannink, N.T., Cocheret de la Morinière, E., and van der Velde, G. (1997). Leaf litter removal by the snail *Terebralia palustris*

(Linnaeus) and sesarmid crabs in an East African mangrove forest. *Journal of Experimental Marine Biology and Ecology*, **215**, 35–48.

Smith, T.J. (1987*a*). Effects of light and intertidal position on seeding survival and growth in tropical tidal forests. *Estuarine, Coastal and Shelf Science*, **110**, 133–46.

Smith, T.J. (1987*b*). Effects of seed predators and light level on the distribution of *Avicennia marina* (Forsk.) Vierh. in tropical, tidal forests. *Estuarine, Coastal and Shelf Science*, **25**, 43–51.

Smith, T.J. (1987*c*). Seed predation in relation to tree dominance and distribution in mangrove forests. *Ecology*, **68**, 266–73.

Smith, T.J. (1992). Forest structure. In *Tropical mangrove ecosystems. Coastal and estuarine studies no. 41* (ed. Robertson, A.I. and Alongi, D.M.), pp. 101–136. American Geophysical Union, Washington D.C.

Smith, T.J., Chan, H.T., McIvor, M.S., and Smith, G. (1989). Comparisons of seed predation in tropical, tidal forests from three continents. *Ecology*, **70**, 146–51.

Smith, T.J., Boto, K.G., Frusher, S.D., and Giddins, R.L. (1991). Keystone species and mangrove forest dynamics: the influence of burrowing by crabs on soil nutrient status and forest productivity. *Estuarine, Coastal and Shelf Science*, **33**, 19–32.

Snedaker, S.C. (1982). Mangrove species zonation: why? *Contributions to the study of halophytes. Tasks for vegetation science*, Vol. 2 (ed. D.N. Sen and K.S. Rajpurohit), pp. 111–25. Junk, The Hague.

Snedaker, S.C., Baquer, S.J., Behr, P.J., and Ahmed, S.I. (1995). Biomass distribution in *Avicennia marina* plants in the Indus River Delta, Pakistan. In *The Arabian Sea: living marine resources and the environment*, (ed. M.F. Thompson and N.M. Tirmizi). pp. 389–94. Balkema, Rotterdam.

Sodhi, N.S., Choo, J.P.S., Lee, B.P.Y.-H., Quek, K.C., and Kara, A.U. (1997). Ecology of a mangrove forest bird community in Singapore. *Raffles Bulletin of Zoology*, **45**, 1–13.

Somerfield, P.J., Gee, J.M., and Aryuthaka, C. (1998). Meiofaunal communities in a Malaysian mangrove forest. *Journal of the Marine Biological Association of the United Kingdom*, **78**, 712–732.

Spalding, M.D. (1997). The global distribution and status of mangrove ecosystems. *Intercoast Network*, March, 20–1.

Spaninks, F. and van Beukering, P. (1997). Economic valuation of mangrove ecosystems: potential and limitations. *CREED Working Paper*, **14**, 1–53. (International Institute for Environment and Development, Amsterdam.)

Stafford-Deitsch, J. (1996). *Mangrove. The forgotten habitat*. Immel, London.

Start, A.N., and Marshall, A.G. (1976). Nectarivorous bats as pollinators of trees in West Malaysia. In *Tropical trees. Variation, breeding and conservation*, (ed. J. Burley and B.T. Styles), pp. 141–50. Academic Press, London.

Steinke, T.D. (1975). Some factors affecting dispersal and establishment of propagules of *Avicennia marina* (Forsk.) Vierh.. In *Proceedings of the international symposium on biology and management of mangroves* (ed. G.E. Walsh, S.C. Snedaker and H.J. Teas), pp. 402–14. Institute of Food and Agricultural Sciences, University of Florida, Gainesville.

Steinke, T.D. and Ward, C.J. (1988). Litter production by mangroves. II. St Lucia and Richards Bay. *South African Journal of Botany*, **54**, 445–54.

Steinke, T.D., Barnabas, A.D., and Somaru, R. (1990). Structural changes and associated microbial activity accompanying decomposition of mangrove leaves in Mgeni Estuary. *South African Journal of Botany*, **56**, 39–49.

Sternberg, L. da S.L. and Swart, P.K. (1987). Utilization of freshwater and ocean water by coastal plants of southern Florida. *Ecology*, **68**, 1898–905.

Sussex, K. (1975). Growth and metabolism of the embryo and attached seedling of the viviparous mangrove, *Rhizophora mangle*. *American Journal of Botany*, **62**, 948–53.

Sutherland, J.P. (1980). Dynamics of the epibenthic community on roots of the mangrove *Rhizophora mangle*, at Bahia de Buche, Venezuela. *Marine Biology*, **58**, 75–84.

Takeda, S., Matsumasa, M., Kikuchi, S., Poovachiranon, S., and Murai, M. (1996). Variation in the branchial formula of semiterrestrial crabs (Decapoda: Brachyura: Grapsidae and Ocypodidae) in relation to physical adaptations to the environment. *Journal of Crustacean Biology*, **16**, 472–86.

Tan, C.G.S. and Ng, P.K.L. (1994). An annotated checklist of mangrove brachyuran crabs from Malaysia and Singapore. *Hydrobiologia*, **285**, 75–84.

Taplin, L.E. and Grigg, G.D. (1981). Salt glands in the tongue of the estuarine crocodile, *Crocodylus porosus*. *Science*, **212**, 1045–7.

Thibodeau, F.R. and Nickerson, N.H. (1986). Differential oxidation of mangrove substrate by *Avicennia germinans* and *Rhizophora mangle*. *American Journal of Botany*, **73**, 512–516.

Thom, B.G. (1984). Coastal landforms and geomorphic processes. In *The mangrove ecosystem: research methods*, (ed. S.C. Snedaker and J.G. Snedaker), pp. 3–17. UNESCO, Paris.

Tilman, D., Knops, J., Wedin. D., Reich, P., Ritchie, M., and Siemann, E. (1997). The influence of functional diversity and composition on ecosystem processes. *Science*, **277**, 1300–2.

Tomlinson, P.B. (1986). *The botany of mangroves*. Cambridge University Press, Cambridge.

Torgerson, T. and Chivas, A.R. (1985). Terrestrial organic carbon in marine sediment: a preliminary balance for a mangrove environment derived from ^{13}C. *Chemical Geology (Isotope Geoscience Section)*, **53**, 379–90.

Turner, R.E. (1977). Intertidal vegetation and commercial yields of penaeid shrimp. *Transactions of the American Fisheries Society*, **106**, 411–6.

Twilley, R.R. (1985). The exchange of organic carbon in basin mangrove forests in a southwest Florida estuary. *Estuarine, Coastal and Shelf Science*, **20**, 543–57.

van der Valk, A.G. and Attiwill, P.M. (1984). Decomposition of leaf and root litter of *Avicennia marina* at Westernport Bay, Victoria, Australia. *Aquatic Botany*, **18**, 205–21.

Vanhove, S., Vincx, M., van Gansbeke, D., Gijselinck, W., and Schram, D. (1992). The meiobenthos of five mangrove vegetation types in Gazi Bay, Kenya. *Hydrobiologia*, **247**, 99–108.

Vannini, M. and Ruwa, R.K. (1994). Vertical migrations in the tree crab *Sesarma leptosoma* (Decapoda, Grapsidae). *Marine Biology*, **118**, 271–8.

Veenakumari, K., Mohanraj, P., and Bandyopadhyay, A.K. (1997). Insect herbivores and their natural enemies in the mangals of the Andaman and Nicobar islands. *Journal of Natural History*, **31**, 1105–26.

Vermeij, G.J. (1974). Molluscs in mangrove swamps: physiognomy, diversity, and regional differences. *Systematic Zoology*, **22**, 609–24.

Voris, H.K. and Jeffries, W.B. (1995). Predation on marine snakes—a case for decapods supported by new observations from Thailand. *Journal of Tropical Ecology*, **11**, 569–76.

Warner, G. (1977). *Biology of crabs*. Elek, London.

Warren, J.H. and Underwood, A.J. (1986). Effects of burrowing crabs on the topography of mangrove swamps in New South Wales. *Journal of Experimental Marine Biology and Ecology*, **102**, 223–35.

Watson, J.G. (1928). Mangrove forests of the Malay Peninsula. *Malayan Forest Records*, **6**, 1–275.

Weissburg, M. (1993). Sex and the single forager—gender-specific energy maximization strategies in fiddler crabs. *Ecology*, **74**, 279–91.

Whitten, A.J. and Damanik, S.J. (1986). Mass defoliation of mangroves in Sumatra, Indonesia. *Biotropica*, **18**, 176.

Wilson, K.A. (1989). Ecology of mangrove crabs—predation, physical factors and refuges. *Bulletin of Marine Science*, **44**, 263–73.

Wolanski, E. (1995). Transport of sediment in mangrove swamps. *Hydrobiologia*, **295**, 31–42.

Wolanski, E. and Chappell, J. (1996). The response of tropical Australian estuaries to a sea level rise. *Journal of Marine Systems*, **7**, 267–79.

Wolcott, D.L. and O'Connor, N.J. (1992). Herbivory in crabs: adaptations and ecological considerations. *American Zoologist*, **32**, 370–81.

Woodroffe, C. (1987). Pacific island mangroves: distribution and environmental settings. *Pacific Science*, **41**, 166–85.

Woodroffe, C. (1992). Mangrove sediments and geomorphology. In *Tropical mangrove ecosystems. Coastal and estuarine studies no. 41* (ed. A.I. Robertson, and D.M.Alongi), pp. 7–41. American Geophysical Union, Washington D.C.

Woodroffe, C.D. (1988). Relict mangrove stand on a last interglacial terrace, Christmas Island, Indian Ocean. *Journal of Tropical Ecology*, **4**, 1–17.

Wyn Jones, R.B. and Storey, R. (1981). Betaines. In *Physiology and biochemistry of drought resistance in plants* (ed. L.G. Paleg and D. Aspinall) pp. 171–204. Academic Press, Sydney.

Yeager, C.P. (1989). Feeding ecology of the Proboscis monkey (*Nasalis larvatus*). *International Journal of Primatology*, **10**, 497–530.

Young, B.M. and Harvey, L.E. (1996). A spatial analysis of the relationship between mangrove (*Avicennia marina* var. *australoasica*) physiognomy and sediment accretion in the Hauraki Plains, New Zealand. *Estuarine, Coastal and Shelf Science*, **42**, 231–46.

Zuberer, D.A. and Silver, W.S. (1975). Mangrove associated nitrogen fixation. In *Proceedings of the international symposium on biology and management of mangroves,* (ed. G.E. Walsh, S.C. Snedaker and H.J. Teas), pp. 643–653. University of Florida, Gainesville.

Glossary

ACEP the Atlantic-Caribbean-East Pacific biogeographic region.

aerenchyma plant parenchyma tissue containing numerous, ramifying air spaces, characteristic of plants living in waterlogged conditions.

aerial roots in mangrove species such as *Rhizophora* these branch off from the trunk at some distance above the soil surface. These supply mechanical support to the tree while avoiding, as far as possible, the anoxic mud. Gas exchange takes place through **lenticels** on the root surface and **aerenchyma** tissue within the underground roots.

anoxic devoid of gaseous oxygen.

ATP adenosine triphosphate.

autotrophic organisms those capable of synthesising complex organic substances from simple inorganic substrates. The principal autotrophic process is of course photosynthesis, and the principal mangrove autotrophs are the trees themselves, and organisms such as algae cyanobacteria on the mud surface.

biomass the total mass of all organisms, or of a specified group of organisms, in a particular habitat, measured in mass per unit area.

DBH (diameter at breast height) a convenient descriptive measure of the size of mangrove trees which, with calibration, can be used to estimate the above-ground biomass.

DOC dissolved organic carbon, usually expressed as $g\ l^{-1}$.

epibenthic organisms those that live at the surface of the substrate.

epibionts organisms that live attached to the surface of other organisms, without depending on them as an energy source.

epifauna the total animal life living on a substrate surface, as opposed to the **infauna**.

epiphytes plants that grow on other plants for support or anchorage, rather than as parasites.

eukaryotes organisms possessing a discrete nucleus surrounded by a nuclear membrane, in contrast to **prokaryotes** such as bacteria and cyanobacteria (blue-green 'algae').

eustatic sea level this reflects the total amount of water in the world's oceans. Actual changes in sea level are affected locally by rising and falling of the land.

GBH (girth, or circumference, at breast height) may be used as an alternative to **DBH**.

heterotrophic organisms those unable to synthesise complex molecules from simple inorganic ones, in contrast to **autotrophs**.

hypersaline water whose salinity is greater than that of the sea (typically around 35‰) is said to be hypersaline.

Indo-Malesia a biogeographic region comprising India, Malaysia, Indonesia, the Philippines and southern China.

Indo-West Pacific the Indian and West Pacific Oceans, which together have faunal similarities which distinguish them from the Atlantic-Caribbean-East Pacific (**ACEP**) region.

infauna the animals that live within a substrate such as mangrove mud.

IWP the **Indo-West Pacific** biogeographical region.

knee roots in mangrove species such as *Xylocarpus* the horizontal underground roots periodically break the surface in the form of knobbly outgrowths (see Fig. 1.3).

lenticel pore in the surface of a tree trunk or root though which gas exchange can take place with the internal tissues.

mangal a term sometimes used to refer to the entire mangrove ecosystem, as opposed to the mangrove trees themselves.

mangrove either a collective term for the trees and shrubs that have evolved adaptations to an intertidal life, such as salt tolerance and resistance to anoxia; or, by extension, to the entire habitat based on such trees.

mangrove associates plants which occur in mangrove habitats, but also elsewhere.

meiofauna the community of small interstitial organisms, often defined pragmatically as those that pass through a 1 mm mesh sieve but are retained by a 0.1 mm mesh. Sometimes other mesh sizes are used as criteria (p. 110).

mM (millimolar) a concentration of 10^{-3} moles l^{-1}.

MPa a pascal (Pa) is a measure of pressure, equivalent to 1 newton m^{-2}.

mycorrhizal fungi these live in association with the root system of plants, and play a major role in nutrient acquisition by their host.

necromass the mass of dead material, such as leaf litter, per unit area.

nitrogen fixation the reduction of atmospheric nitrogen to ammonia, usually carried out by nitrogen-fixing bacteria.

osmoconformer an organisms whose internal body fluids conform with the osmotic concentration of the external medium, in contrast to an **osmoregulator** which spends energy maintaining a difference between internal and external media.

pneumatophores vertical, spike-like extensions of the underground roots of mangroves such as *Avicennia* and *Sonneratia*. Because of their abundance of lenticels and aerenchyma, they allow gas exchange to take place. Pneumatophore growth, at least in *Avicennia*, is facultative, and more pneumatophores develop in more anoxic soils.

POC particulate organic carbon, in contrast to **DOC**. The distinction is useful, although there is probably a spectrum from true solution, through colloidal suspension, to discrete particles.

prokaryotes organisms, such as bacteria and cyanobacteria (blue-green 'algae') in which the genetic material is not contained within a discrete nucleus with a nuclear membrane.

redox potential (or oxidation-reduction potential) a measure of the tendency to act as an oxidising (electron-acceptor) or reducing (electron-donor) agent. Redox potential is measured in millivolts (mV): increasingly anoxic soils have more negative redox potentials.

sesarmine crabs numerous, very similar species of the subfamily Sesarminae, family Grapsidae, which seem to have evolved and diversified largely for the purpose of confusing taxonomists. They are important constituents of the mangrove fauna (p. 81).

stable isotopes carbon, nitrogen and oxygen exist in a number of stable isotopic forms. When, for example, carbon is absorbed by mangroves in the form of CO_2, the two principal stable isotopes (^{12}C and ^{13}C) are assimilated differentially, giving a characteristic ^{12}C: ^{13}C ratio in the tissues which differs from the ratio found in, for example, algae. The characteristic stable isotope 'signature' of mangroves is reflected in the tissues of species that feed predominantly on mangrove material so that stable isotope ratios can be used to trace the passage of mangrove carbon through food chains.

stilt roots an alternative term for **aerial roots**.

stipule new leaves form as a bud enclosed in a pair of stipules, which are shed as the bud expands. Stipule fall correlates closely with the rate of appearance of new leaves, hence it may be taken as an index of productivity.

xylem the main water-conducting tissue of vascular plants.

Index

abscisic acid (ABA) 24
Acanthus 14, 23, 26, 44
accretion of shore 42, 43–4, 47–8, 183
  see also sedimentation
ACEP *see* Atlantic-Caribbean-East Pacific
  region (ACEP)
Acrostichum 2, 39, 44, 46, 105, 169, 175–6
  A. aureum 154–5
  A. speciosum 53
Adinia 129
Aëdes 59
  A. alternans 59
  A. amesii 59
  A. pembaensis 59
  A. vigilax 59
Aegialitis 2
Aegiceras 2, 12, 13, 14, 16, 19, 23, 27, 38, 50,
  54, 62, 91, 192
  A. corniculatum 53, 88–9
aerenchyma 5, 7–8, 128
Africa 71
  East 3, 59, 88, 93, 154–5
  North 158
  South 122
  West 3, 74, 107, 154–5, 158, 162–3
agar 77
agouti (*Dasyprocta*) 73
alcohol dehydrogenase 11
alcohol production 11, 172, 173
algae 48, 77–8, 83, 85–6, 95, 112, 138–9
  response to nutrient enrichment 181, 185, 193
alligator 67, 130
alpha diversity 152
Alpheus 105, 114
Amazon 33, 130
Ambassis 142
America, Central 28, 29, 69, 73
  North 158
  South 3, 57, 73, 153, 159
ammonia, ammonium 19, 20, 21, 23, 50, 106,
  115, 136
amphibians 63–4
amphipod 58, 111, 114, 129
Amyema thalassium 52
Anadara granosa (blood cockle) 177
anaerobic conditions 45, 8–11, 17–18, 20, 32,
  49, 100–101, 149

anchorage by roots 11, 79
anchovy 113
Andaman Islands 54
annelids 79
  see also polychaetes, oligochaetes,
    archiannelids
Annobon 162–3
anoxia, *see* anaerobic conditions, redox potential
ant plants, ant house plants 58
antelope 73
ants 57–9, 61, 165
Apis mellifera 62
Apocrytes 115
aquaculture 174, 177–80, 196–7
Arabia 73
  see also Sinai
Aratus pisonii 82, 84, 93
archerfish 113
archiannelids 111
Arecaceae 157
Argyrodes 62
Arius 142
arthropods 1647
  see also crustacea, insects, spiders
ascidians 79
Atlantic Ocean 79, 81, 155, 156
  see also Atlantic-Caribbean-East Pacific region
Atlantic-Caribbean-East Pacific region
  (ACEP) 3, 152–5, 160–1, 169
ATP 10
Atta 58
Australasia 3, 153–4, 155
Australia 3, 18, 35–7, 48, 49–50, 52, 56, 57,
  58, 59, 65, 71, 73, 74, 75, 88, 90, 92, 95, 120,
  121, 127–8, 130–1, 154–6, 158, 173
  see also Darwin, Queensland, Hinchinbrook
  Island
Avicennia 2, 8, 10–11, 12–16, 18, 23, 26–7,
  28–31, 36, 38–9, 44, 48, 50, 52, 54, 55, 56,
  62, 69, 74, 77, 78, 82, 111, 113, 120, 122,
  124, 128, 154, 158, 164, 184–5, 192
  A. alba 37, 164
  A. bicolor 29, 38
  A. germinans 19, 28–9, 38, 92, 163, 184
  A. intermedia 37
  A. marina 4, 7, 10, 15, 28, 30, 40, 44, 53,
    55–6, 889, 92, 119, 134, 54, 164, 187, 192

A. officinalis 164
Avicenniaceae 2
  see also Avicennia
axis deer (*Axis axis*) 73

bacteria 4–5, 19–20, 23, 48, 74, 78, 94, 95,
  111, 125–6, 128–9, 131, 137, 140
Bahamas 137, 163
Bahía las Minas, Panama 182–3
Balanus 78
bandicoots 73
Bangladesh, *see* Sundarbans
barnacles 78, 107, 185
barriers, biogeographical 152–5, 158–60
bats 24, 74–6
  pollination by 75–6
Bay of Bengal 33–42
bee-eaters 174
bees 23–4, 62, 165, 173
beetles 54, 69
  see also fireflies, *Coccotrypes*
Belize 54, 80, 92
beta diversity 152
biodiversity 45–6, 148, 151–69, 170
  and ecosystem function, *see* ecosystem
    function
  local patterns of 160–3
  loss of 192
  of mangrove fauna 1, 51, 155, 164–7
  of mangrove plants 151–63
  of mangrove root fauna 79–80
  microbial 169
  regional patterns 152–60
biogeographical regions 152–5, 159–61, 168
biogeography 38, 45–6
  of mangrove root fauna 79–80
Bioko 162–3
biomass 50, 56, 74, 137, 168
  of crabs 126–7
  of trees 117–24, 131–2
birds 234, 54, 6872, 198
  diet of 68–72
  see also particular group, e.g. passerine,
    kingfisher, stork
bivalves 111, 142, 173
  see also cockles, mussels, oysters, piddocks,
    shipworms
bluegreen 'algae' 48, 77, 107
Boiga dendrophila 65–6
Boleophthalmus 14–15
Bombacaceae 2
  see also Camptostemon
boring organisms, *see* wood-borers
Borneo 74, 173
Bostrychia 77–8
bostrychietum 77
Brachydontes 129
Brachyura, *see* crabs
Brahmaputra 33, 36, 42
Brazil 105, 158
bromeliads 51

Bruguiera 2, 6, 8, 14, 23, 27, 36, 44, 52, 57, 74,
  111, 127, 158, 156
  B.cylindrica 37
  B. exaristata 56
  B. gymnorrhiza 37, 39, 40, 44, 53, 56, 89, 93
  B. parviflora 14, 37, 39, 44, 56
  B. sexangula 37, 56
bryozoans 79
Bubulculus ibis 68
buds 93
buffalo, water (*Bubalus*) 73, 192
bugs (Hemiptera) 54–5
  see also coccids
burrows 21, 50, 83, 89, 101–2, 106, 114–15
butterflies 23, 58, 61–2

C:N ratio, *see* carbon: nitrogen ratio
cadmium 180
Caesalpina 51
Caiman crocodylus, caiman 66, 130
Calamus erinaceus 51
California 154
Callinectes sapidus 84
Caloglossa 77
camels 73, 192
Cameroon 162–3
Camptostemon 2
canopy 44–6, 104
  see also gaps
carbohydrates 87
carbon assimilation, *see* photosynthesis
carbon dioxide 5, 8–10, 17, 141, 171
carbon flux 127, 128–9, 131–2, 136–42
  stable isotopes 85–6, 95, 107–8, 109,
    137–41, 143, 146–8
carbon: nitrogen ratio 54, 87–9, 126
carbonate settings 35, 189
Cardisoma 83
  C. carnifex 89
Caribbean 35, 56, 79, 81, 92, 158, 181, 183,
  189
  see also Atlantic-Caribbean-East Pacific region
Caroni Swamp, Trinidad 68, 174
cat, fish (*Felis viverrina*) 73
Catanella 77
catch per unit effort (CPUE) 193–4
catfish 113
cattle 192
cellulase 83
cellulose 78, 125–6, 128
Ceratopogonidae 59
Cerberus 68
Cercopithecus 74
Ceriops 2, 12, 14, 24, 27, 30, 36, 57, 58, 69, 87–8,
  111, 124, 156, 192
  C. australis 40, 56
  C. candoleana 39
  C. decandra 14
  C. tagal 4, 37, 53, 56, 87, 89, 93
Cerithidea 107, 109, 112
Cervus duvauceli 73

Cetacea 73
CFCs 187
Chaerophon jobensis 75
Chalinolobus gouldii 75
charcoal 172
    see also Matang
China 53, 556, 88, 174, 177, 179, 180
Chiromanthes bidens 88–9
    C. maipoensis 88–9
chloride ions, *see* salt
chlorofluorocarbons (CFCs) 187
chlorophyll 8, 26, 121, 163
Chlorophyta 78
Christmas Island 36, 42
Cleistocoeloma merguiensis 84, 99
Clibanarius 104
    C. panamensis 78
climate 43, 80
    change 71, 185–9
C:N ratio, *see* carbon:nitrogen ratio
CO_2, *see* carbon dioxide
coastal accretion, erosion, *see* accretion, erosion
cobra, king (*Ophiophagus hannah*) 65
coccids, scale insects 54, 57, 61
Coccotrypes 56
cockle 177
Combretaceae 2
    see also Laguncularia, Lumnitzera
community structure 182, 185
competition 40–1, 43, 80, 89, 112, 160–2, 168
Congeria 129
Conocarpus 57–8
convergent evolution 2, 13, 17, 30, 52
Cook, Captain 57–8
copepods 110–12
copper 180
Coral Creek 136
coral reefs 35, 141–2
cormorants 68
Corner Inlet, Australia 3
Costa Rica 26
cotyledon 24–6
CPUE, *see* catch per unit effort
crab burrows 59, 62
    exclusion experiments 90, 112
    feeding 82, 83–97, 104, 123–4, 126–7,131
    osmoregulation 102–3
    predators 66, 104
    stress 104
    water relations 93, 99, 102–4
    zonation 99
crabs 50, 142, 143, 174, 196, *see also* hermit
    crab, Portunidae, Grapsidae, Ocypodidae,
    Xanthidae
    allocation of time by 90, 99, 101, 104
    as prey 66, 69, 74, 93, 97, 127, 131
    diet of 83–93, 94–8, 104,169
        see also predation
    effects on mangroves of 105–6
    fiddler, *see* Uca
    predating mangrove propagules 90–2

temperature tolerance in 101–2
    tolerance of anoxia by 100–1
    tree-climbing 92–3
Crassostrea 79
Cretaceous 157–8
crocodiles 66–8, 130
Crocodylus acutus 66–7
    C. niloticus 66
    C. porosus 66, 130
Crustacea, *see* amphipod, copepod, crab,
    cumacean, isopod, shrimp
cryptovivipary 26–7
Cumaceans 111
currents 143–4, 147
Cyanobacteria (bluegreen bacteria) 48, 77, 117
cyclone, *see* hurricane
Cyprinodon 129

dams, effects of 33–4, 190–2
Darwin, Australia 69–70
Dasyprocta 73
DBH, *see* diameter at breast height
decay, *see* decomposition
decomposition 19, 20, 21, 87–90, 92, 123–6,
    131, 135, 169, 188
deer 73
defoliation 556, 164, 179, 181, 183
deposit feeding 94–7, 105, 107–8, 114, 131
Derris 44, 51
desalination by mangroves 12–13
desiccation, resistance to 1, 24, 78, 109, 114–15
Desulfovibrio 19
detritus, detritivores 107–8, 113, 129–31,
    7–40, 141, 147–8
    see also crab diet, leaf litter
diameter at breast height (DBH) 117–19
diatoms 77, 93, 94, 95, 129
Dicaeum hirundinacaeum 52
Dicyathifer 109
dispersal 153, 156, 159, 162–3, 164–6
dissolved organic carbon (DOC) 126, 129, 131,
    136, 137, 140, 142
dissolved organic matter (DOM), *see* dissolved
    organic carbon
disturbance 43–5, 80
    see also storms, hurricanes
diversity, *see* biodiversity
    genetic 152, 163–4
    microbial 169
DNA 24, 164
DOC, *see* dissolved organic carbon
dolphins 73
DOM *see* dissolved organic carbon
Drosophilididae 24
dugong 73
durian (*Durio zibethinus*) 76
dwarf mangroves, *see* stunting

economic value of mangroves 195–8
    see also aquaculture, exploitation, fisheries
ecosystem engineers 105–6

function 116–17, 130–2, 166–9
  modelling 116–32
  stability 166–9
ecotourism 174, 177, 195, 197, 198
Ecuador 185
effluent 180–1, 190, 193, 196, 197, 198
egrets 21, 68
Egretta thula 68
Egypt 10
  see also Sinai
energy flow through ecosystem 49, 120–32
England 158
Enteromorpha 78
environmental settings of mangroves 33–6
enzyme function 11, 13, 19
Eocene 157, 158, 159
Eonycteris spelea 75–6
epibionts 19–20, 78–80
epiphytes 1, 19–20, 46, 51–2, 58, 77–8
Episesarma 105
  E. versicolor 84, 92–3, 99, 127
epizoites 19–20
erosion, coastal 42, 43–4, 150
  mangroves as defence against 174, 179, 193,
    196, 197
estuaries 33–5, 36–7, 39–40, 42–3, 45, 47, 138,
  143–4, 160–1
estuary, salt wedge 143–4
ethanol 11
  see also alcohol
Eudocimus ruber 68, 174
Euphorbiaceae 2
  see also Excoecaria
Europe 158–9
eustatic sea level rise 42, 188
evolution of mangroves 2, 156–60, 164
evolution, convergent 2, 13, 17, 30, 52
Excoecaria 2, 13–14, 24, 27, 52, 54, 55
  E. agallocha 53, 54
exploitation of mangroves 107, 170–80
  see also aquaculture, firewood
export of organic matter 123–4, 131, 136–42
  see also outwelling
extinction 158, 161

faeces 97, 127, 131
fantails 69–70
feeding guilds 69–71
Felis viverrina 73
fern 2, 51
  see also Acrostichum
Fernando Po 162–3
Fiji 66
filter feeding 78, 129, 147
  see also barnacles, bivalves
fireflies 5961, 174
firewood 172, 192
fish 112–15, 129, 142–3, 148
  diet of 142
fisheries 130, 144–9, 150, 170, 172, 173–4,
  177, 193–4, 196, 197

flatworms 111
  see also platyhelminths, turbellarians
flavoglycans, flavolan 87
flies 24
flocculation 47
Florida 5, 12, 19, 21, 22, 35, 37, 43, 66, 80, 92,
  120, 122, 132, 137, 138, 141, 164–7, 184
flowers, flowering 234, 278, 46, 62, 701
  see also pollination
Fly River, Papua New Guinea 33, 47
flycatcher 69–70
flying foxes 65, 75
fodder 172
fossil history of mangroves 156–8
France 158
fringing mangrove forests 171
frog, crab-eating 63–5
fruit 76
functional groups 79, 116–17, 130–1, 167–8
fungi 58, 125–6
Fusarium 126

gamma diversity 152
Ganges 33, 36, 42, 171
gaps in forest 30, 32, 44–5, 90–1, 184
gas transport in roots, *see* roots, aeration of
Gasteracanthidae 62
gastropods 68, 79, 85–6, 92, 106–9, 142
  see also Cerithidea, Littoraria, Morula, Telescopium,
    Terebralia, Thais
  shell colour 108
  deposit feeding 106–7
  predatory 78, 79, 107
gastrotrichs 110, 111
GBH, *see* girth at breast height (GBH)
Gecarcinidae 83
Gecarcoidea 83
gei wai system 177–8, 196
genetic diversity 152, 163–4
geographical distribution of mangrove
  species 2–3, 31–2, 35, 36, 107, 151–63,
  187–8
geomorphological change 38, 42–3
Germany 158
germination 24–6
gerygone, *Gerygone* spp. 69–70
girth at breast height (GBH) 117, 119
global climate change 171, 185–9
global warming 171, 187–8
glycan 87
glycinebetaine 13
glycolysis 10–11
goats 192
gobies 113
Goniopsis 81, 92
goods and services of mangroves, *see* economic
  value
Gracilaria 77
gradients, environmental 34, 38, 39–42, 80
Grapsidae 81–3, 92, 99

see also Aratus, Goniopsis, Metopograpsus, Sesarma,
  sesarmine etc.
grass 149
great tit (*Parus major*) 71
greenhouse effect 187
guano 21, 68

haemoglobin 64
Haliclona 19–20
harpacticoid copepods 110–12
haustorium 52
Hawaii 153
hawk, crab-eating 71
heavy metals 180, 193
herbivory 52, 149
  crab 83–93
  insect 5, 27, 53–7, 63, 73, 74
  mollusc 7
Heritiera 2, 42, 54, 57
  H. littoralis 53, 56
hermit crab 78, 104
herons 68, 97
Herpestes 73
Hinchinbrook Island, Queensland 21–3,
  111–12, 136
Holocene 35
Honduras 174
honeyeaters (Meliphagidae) 70–1
Hong Kong 55–6, 88
Hormosira 77
hummingbirds (Trochilidae) 69, 71
hurricanes and typhoons 28, 132, 169, 171
Hydnophytum formicarium 58
hydroids, hydrozoa 79, 111
hydrological regime, altered 171
  see also dams
Hydrophidae 65
Hymenoptera, *see* ants, bees, wasps
hypersalinity 12, 14, 16, 18, 35, 39–40, 49, 195
  see also salinity
Hypochrysops 58
hypocotyl 24–6

ibis, scarlet 68, 174
Ilyoplax 9
impacts of human activities, *see* aquaculture,
  clearance, exploitation, management,
  pollution
inbreeding 163
India 60, 107, 164
  see also Sundarbans
Indian Ocean 152, 158
Indo-Malesian region 3, 154–5
Indonesia 55, 60, 178, 178
Indo-Pacific 57
Indo-West Pacific (IWP) region 3, 79, 105,
  152–5, 159–61, 168, 169
Indus Delta 4, 12, 18, 33, 35, 189–96
insect attack, *see* herbivory
insects 53–62, 114, 164–6
  see also ants, bees, beetles, butterflies etc.

interactions between environmental
  variables 14, 161–8, 32
interactions between mangrove and other
  ecosystems 133–50
interactions between species 38, 435
  see also competition, succession, symbiosis
Intsia 37
ion uptake 16
  see also nitrate, phosphate, salt
Iridomyrmex 58
islands 153, 161–3, 164–6
Isoodon 73
isopod 78–9
isotopes, stable, *see* carbon, oxygen
IWP, *see* Indo-West Pacific

Java 178
jellyfish 174
Jiulong Jiang, China 174

Kandelia 2, 24, 27, 30, 156
Kandelia candel 8–9
Kenya 105, 141, 143
Kimberley, Australia 75
kingfishers 68–9, 97, 174
kinorhynchs 110–11
Klang Straits, Malaysia 39, 145, 147
kleptoparasitism 62
Krebs cycle 10
K-selection 45
Kuwait 188

lactate 10–11, 100
Laguncularia 2, 14, 19, 27, 31, 57, 155, 188
Laguncularia racemosa 19, 29, 38, 92, 163
Lampyridae (fireflies) 54, 59–61, 174
land crab, *see Cardisoma*, Gecarcinidae,
  Gecarcoidea, Ucides
langur monkey (*Presbytis*) 74
larvae 79, 80, 98, 113, 127, 143–4, 147, 177–80
Laticauda colubrina 65
latitude, effects of 119–20, 122, 148, 149, 159
lead 180
leaf composition 46, 54, 87–9
  see also carbon: nitrogen ratio, tannin
leaf damage by insects, *see* insect herbivory
leaf litter 85–90, 92, 107–8, 112, 121–9, 131
  see also crab feeding, detritus, litter
leaf palatability 87
leaf production 22–3
leaf structure 11–12, 17, 19, 46, 54
leafcutter ants (*Atta*) 58
leaf-eating 124, 126–7, 131
  see also crab feeding, insect herbivory
leaves 138–9, 141
Leguminosae 44, 51
lenticels 8–10, 20, 78
Lepidoptera 58
  see also butterflies, moths
lianes 51
lichens 51

light 17, 31, 44, 46
  attenuation through canopy 120–1
  effects of 108
  see also photosynthesis
lignin 137
lignocellulose 128
litter production 22–3, 35–49, 183
  fate of 123–9, 131, 138–40, 169
  see also crab feeding, decomposition, leaf litter
Littoraria 107–8
  distribution within mangal 108
  L. angulifera 108
  L. intermedia 108
  L. pallescens 108
  L. scabra 108
Liza 113, 142
lizards 64, 68
  see also Varanus
Lizashtonia hirsutisoma 110
Lophogobius 129
Loranthaceae 52
loss of mangroves 170–1, 192–3
  see also shrimp farming
Lumnitzera 2, 19, 27, 69, 155
  L. littorea 39, 53
Lythraceae 2

Macaca, macaque monkey 74
Macroglossus minimus 75
Macrophthalmus 99
  M. setosus 103
Mai Po 55–6, 88
maize 16
malaria 59
Malaysia 7, 33, 36–7, 39, 51, 53, 57, 60, 62,
  69, 71, 74–6, 77, 90, 92, 99, 112–13, 120,
  125, 127, 129, 138–9, 142, 145–7, 148, 155,
  171, 173, 174, 196
  see also Matang
mammals 72–6
  see also bandicoots, deer, monkey, etc.
management 132, 168–9, 170–80, 185,
  189–95, 198–200
mangrove associates 1, 51–2
  distribution, affected by crabs 90
  habitat 1, 3–23, 34–6
  reproduction 23–32, 46
    see also flowers, pollination, propagules
  species, distribution on shore, *see* zonation
  taxonomy 12
mangroves as nursery areas 142, 146, 147, 148
  basin 345
  effects on the environment 10, 43–4, 47–50.
  hammock 35
  inland 35–6
  overwash 35
  riverine 33–5, 47, 74, 138
  river flow, estuaries
  scrub 35
  tide-dominated 34
mannitol 13

mantis shrimps 104
Maryland 158
Matang forest, Malaysia 33, 44, 118, 121, 125,
  168–9, 175–7
Maximum Sustainable Yield (MSY) 145
medicines 172, 173
Mediterranean 158
Megachiroptera, *see* bats
meiofauna 95, 110–12, 126, 128
Meliaceae 2
  see also Xylocarpus
Meliphagidae 70–1
Melita 129
mercury 180
Merguia 105
Metapenaeus mutatus 146–7
Metaplax elegans 99
methane 5, 187
Metopograpsus 81–2
microalgae 117
microbial action 23, 123–9, 181
  diversity 169
  see also bacteria, fungi
Microchiroptera *see* bats
midges, biting (Ceratopogonidae) 59
mimicry 2
Miocene 157
mistletoe 52
mistletoe bird 52
mitochondria 13
molluscs 155, 157, 174, 196
  see also bivalves, gastropods
mongoose (*Herpestes*) 73
monitor lizard (*Varanus indicus*) 66, 174
monkeys 57, 74, 97
Mormopterus loriae 75
Morula lugubris 78
mosquitoes 59, 179
moths 23, 55–6, 62
Mozambique 37
mud crab (*Scylla*) 107, 174, 179
mud lobster 44, 105
mudflats 43–4, 68, 109
mudskippers 59, 66, 68, 97, 113–15
mullet (*Liza*) 113, 142
Murrayella 77
mussel 109, 129
mycorrhizal fungi 20
Mycteria cinerea 177
Myrmecodia 58
myrmecophytes 58
Myrsinaceae 2
  see also Aegiceras
Myrtaceae 2
  see also Osbornia

Nasalis larvatus 74
Nasutitermes 57
necromass 85, 123–4
nectar 13, 23–4, 58, 71, 75
nematodes 95, 110, 111, 112

nemerteans 114
Neosarmatium meinerti 59, 84
    N. smithi 82, 87–8
Nephila clavipes (golden silk spider) 62
Nerodia fasciata 66
New Caledonia 153
New Guinea 52, 60, 71, 108, 153–4, 158
New Zealand 3, 154, 185
niche width 160–1
niches, feeding 71–2
Nicobar Islands 54
Nigeria 105
nitrate 4, 19, 20–1, 68, 72, 135, 136
nitrite 20, 136
nitrogen 4, 19–21, 22, 23, 54, 58, 116, 127,
    137, 179
    see also ammonia, nitrate, nitrite
nitrogen fixation 19–21, 77, 78, 136, 137
nitrous oxide (N$_2$O) 20, 23
noctuid moths 55
Nothopterix 55
NPP, Net Primary Production, *see* production,
    primary
nutrients, inorganic 50, 56, 58, 116, 135, 141,
    149
    see also phosphorus, nitrogen
Nycticeius greyi 75
Nyctophilus arnhemensis 75
Nypa 2, 24, 27, 42, 158, 159
    N. fruticans 154, 173

Ocypodidae 99, 103
    see also Ilyoplax, Macrophthalmus, Uca, Ucides
Odocoileus 73
Oecophylla smaragdina 57
oil pollution 181–3
Oligocene 157
oligochaetes 111, 112
Ophiophagus hannah 65
Ophiusa 55
Orcaella brevirostris 73
orchid 51
Osbornia 2, 14, 27
osmoregulation, amphibian 63–5
    crabs 102–3
    reptile 66–8
osprey 68
otter 73, 93, 177
outwelling 34, 135–42, 146, 147
oxygen 4, 8–10
    debt 11, 100
    stable isotopes 12
    see also anaerobic conditions
oysters 79, 129, 158, 185

Pacific 153, 156, 188, 189
    see also Atlantic-Caribbean-East Pacific,
        Indo-Pacific, Indo-West Pacific
Pakistan, *see* Indus delta
Palaemonetes 129
Palaeocene 157–8

Palaeowetherellia 158
Palmae, palm trees 51
    see also Nypa
Panama 38, 69–71, 105, 153, 154, 158
Panthera tigris (Bengal tiger) 73, 74
Papua New Guinea 52, 60, 71, 108, 153–4,
    158
Paracleistostoma mcneilli 103
Parasesarma plicata 82
parasitic plants 52
Pardosa 62
particulate organic carbon (POC), particulate
    organic matter (POM) 126, 129, 131, 136,
    141–2, 143
Parus major (great tit) 71
passerines 71
patch dynamics 44–5
pelican 21, 68
Pelliciera 2, 17, 31, 158
    P. rhizophorae 29, 38
Pellicieraceae 2, 157
    see also Pelliciera
Pemphis 2
Penaeidae 145–8
    diet 147–8
    see also aquaculture, shrimp
Penaeus 177
    P. merguiensis 146–7
Perameles 73
pericarp 24, 26, 28, 30
Periophthalmodon 114–15
Periophthalmus 114–15
    P. cantonensis 114
    P. sobrinus 115
periwinkles 107
pesticides 181
pH, soil 42, 109
Philippines 60, 113, 158, 171, 178, 185
Pholadidae 78
phosphate 5, 19, 21–3, 68, 123, 135
phosphorus 116, 137, 179
    see also phosphate
photosynthesis 1, 8, 16–17, 77, 120–1, 186, 187
    measurement of 120–1
physiological stress 104, 180
piddock 78
pigs, wild 73
pioneer species 43–6
Pipistrellus tenuis 75
pistol shrimps (Alpheidae) 105, 114
plankton 47, 59, 80, 98, 138, 143–4, 147
Platanista gangetica 73
platyhelminths 111
Pliocene 157
Plumbaginaceae 2
pneumatophores 4, 6–8, 10–11, 77–8, 82, 113,
    148, 181
POC, *see* particulate organic carbon
pollen, fossil 157–8
pollen, pollination 23–4, 31, 46, 62, 71, 75–6,
    163–4

pollution, effects on mangroves 180–3, 193, 195, 198
polychaete worms 79
polymorphism 108, 164
Polyrachis sokolova 58–9
POM, *see* particulate organic carbon
Pomadasys 142
porpoises 73
Portunidae (swimming crabs) 104
 see also Scylla
prawn, *see* shrimp
predation 78–80, 98, 104, 108, 111, 112, 130–1, 148
 by crabs 79, 92, 104, 107
 of crabs 90, 93, 97, 104
 by fish 81, 97, 114
 by molluscs 106
Presbytis (langur monkey) 74
Principé 162–3
Prionos ornata 110
proboscis monkey (*Nasalis*) 74
Procyon cancrivorus 73
production, bacterial 128
 primary 21–3, 26–7, 49, 77, 106, 120–3, 131, 140, 145, 149, 55–6, 166–9
 secondary 97, 126–7
 see also photosynthesis
productivity, *see* production
prokaryotes 77
 see also bacteria, Cyanobacteria
proline 13
propagules 24–32, 38–9, 40, 43, 46, 55–7, 74, 122, 162–3, 172, 185, 193–5
 dispersal 24, 27–32, 38–9, 46, 56–7, 162–3
 mortality 27–8, 56, 77, 90–2, 185
 settlement 27–32, 38–9
 survival 43
protozoa 95, 110, 111, 112
Pseudoapocrytes 115
pseudofaecal pellets 94–5, 97, 112
Pteridaceae 2
 see also Acrostichum
Pteroptyx cribellata 60
 P. malaccae 60
 P. tener 60
Pteropus 75
Puerto Rico 113, 184
pyralid moths 55
python 65

Quaternary 157
Queensland, Australia 53, 72, 87, 113, 123–4, 136

racoon 97
racoon (*Procyon cancrivorus*) 73
radula 107, 108
rainfall 19, 34, 183, 187, 190
rainforest 1, 45–6, 71–2
Rana cancrivora 63–5
rat 74

rattan 51
Red Sea 10–11, 134, 153
redox potential 4–5, 11, 23
replanting of mangroves 175, 185, 193–5, 198–9
reptiles 64–8
respiration 4–11, 16, 31, 81, 109, 115, 121, 132, 141
Rhinoceros sundaicus (Java rhinoceros) 3
Rhipidura rufiventris 72
Rhipidura sp. 72
Rhithropanopeus 129
Rhizophora 2, 5–11, 12–14, 17, 21–3, 24–7, 30–2, 35, 36, 38, 44, 54, 56, 58, 72, 74, 78, 80, 92, 111, 117–18, 120, 121, 124, 125, 128, 155, 156, 158, 163, 164, 176, 182, 184, 185
 R. apiculata 27, 37, 39, 44, 53, 56, 17–7, 194
 R. harrisoni 29, 38, 162–3
 R. mangle 5, 15, 19, 21, 26, 27, 29, 38, 43, 78, 92, 153, 163, 164, 184
 R. mucronata 37, 39, 40, 89, 93, 193
 R. racemosa 163
 R. samoensis 153
 R. stylosa 40, 53, 56
 R. X lamarcki 53
Rhizophoraceae 2, 24, 156–7
 see also Bruguiera, Ceriops, Kandelia, Rhizophora
Rhodophyta 77
rias 35
river diversion, *see* dams
river flow, effects of 19, 33–6, 43, 47, 136, 138, 143–4, 161
 see also dams, estuaries, riverine mangroves
robin, mangrove 70
rodents 73, 74
root communities 77–80
root fauna 19–20
root: shoot ratio 16
roots 4, 8
roots and nutrient uptake 19–21
 and salt 11–13, 14
 aeration of 5–11, 78
 aerial 5–8, 48–9, 77–80, 93, 104, 113, 117–18, 120, 181–3
 effects on sedimentation 48
 prop 10–11, 158
 stilt 5
 structure 5–8, 10–11, 16, 17
 underground 4–11, 49, 120, 123, 128, 150
r-selection 45
Rubiaceae 2

sabellid worms 79
salinity 35, 41–50, 143–4
 effects of mangroves on 49–50
 gradients 39–42, 47, 160–1, 188
 optimal 4–15, 39–41
 variation, response to 18, 30, 32
salt accumulation 13, 14, 49
 exclusion 12–13, 14, 47, 50
 glands, plant 13–14

glands, reptiles 67–8
secretion 13–14, 47, 54
tolerance 11–18, 26, 30, 32, 39–41, 51, 52, 102–3, 109
saltflat 49
salt marsh 1, 97, 106, 135, 144, 145, 149–50, 189
Samoa 153
sandflies 59
Saõ Tomé 162–3
saponins 54
saprophytes 58
Sarawak 173
scale insects (Coccidae) 54, 57, 61
scale, importance of 36, 42–3, 80, 168
Scartelaos 114–15
scolytid beetles 56
Scylla 107, 174, 179
   S. serrata 104
Scyphiphora 2, 17
   S. hydrophyllacea 54
sea level rise 35, 42–3, 171
seagrass 30, 138, 140–3, 145
season, effects of 122
sediment 33–6, 49–50, 138–41, 149, 180–1
   bacteria 128–9
   trapping 34, 43, 47–8, 141
sedimentation 33, 47–8, 185, 189, 190–2
seedlings 7, 12–13, 14–15, 24–32, 55, 183, 184, 192, 193
   predation of 90–2
   see also propagules
seeds 52, 57
Selangor river, Malaysia 60, 112–13, 174
Senegal 74
serpulid worms 79
Sesarma 63, 81–3, 102, 104
   S. eumolpe 84
   S. leptosoma 93
   S. onychophora 84, 99, 127
   S. plicata 84
   S. reticulatum 101
   diet 84
   geographical distribution of 81–3
   see also Chiromanthes
sesarmine crabs (Sesarminae) 81–2, 114, 116, 127, 149
   abundance 127
   diversity 155
   feeding 83–93, 124, 126–8, 131
   geographical distribution 155
   respiration 81, 100–1
   temperature control 101–2
   see also Aratus, Chiromanthes, Episesarma, Neosarmatium, Parasesarma, Sesarma, Sesarmoides
Sesarmoides kraussi 84
shade 17, 30, 31–2, 44, 46, 56, 78, 109
sheep 192
shipworms (Teredinidae) 78, 109, 128
shrimp 129, 144–8

farming 174, 177–80, 196–7
   dependence on mangroves 144–8
Sinai 10–11, 35
Singapore 61, 69
sipunculans 113
snails, *see* gastropods
snake, cat 65–6
snakes 64–8, 97
   osmoregulation 67–8
   sea 62
Society Islands 153
Sonnerataceae 2, 157
Sonneratia 2, 8, 13, 14, 23, 27, 39, 42, 57, 74, 76, 111, 125
   S. alba 17, 37, 75, 89
   S. caseolaris 61, 75
   S. griffithi 37
   S. lanceolata 14, 17
   S. ovata 76
South-east Asia 45, 69, 73, 74, 156, 158, 171
   see also under country
Spain 158
Spartina 106, 149
species diversity, *see* biodiversity
species interactions, *see* ant-house plants, competition, symbiosis
species-area relationship 152, 161–2, 166–7
Sphaeroma 78–9
spiders 62
sponges 19–20, 79
Squilla choprai 104
Sri Lanka 148, 158
Sterculiaceae 2
   see also Heritiera
stipules 22–3
stomata 17, 186
Stomatopoda 104
stork, milky (*Mycteria cinerea*) 177
storks 68, 177
storms 43. 80
   see also hurricanes
stress, environmental 104, 180
   see also anaerobic conditions, desiccation, salinity, temperature,
stunting 8, 35, 37, 120, 122, 195
succession 38, 43–5, 47, 48, 80, 112, 132, 184–5
   microbial 126
sucrose 173
Sulawesi 178
sulphate 5
sulphide 5, 11, 106
sunbirds (Nectariniidae) 71
Sundarbans 33, 42, 73, 130, 171–2
Surinam 71
sustainable management, *see* management
symbiosis 58, 78, 128

Tachyurus 142
tadpoles 63–4
tailor ant 57

Taiwan 179
tannin 54, 87–9, 112, 125–6, 140, 173
Tanzania 37
Taphozous flaviventris 75
 T. georgianus 75
tardigrades 110–11
tectonic movement 42, 159–60, 188
Tedania 19–20
Telescopium 107, 173
temperature tolerance 2–3, 101–2, 108–9, 145,
 149, 152, 159, 180, 184, 187–8
temperature, global, rise in, *see* global warming
Terebralia 107, 112
 T. palustris 107–8
Teredinidae 78, 109, 128
termites 57, 128
Tertiary 157
Tethys Sea 158–9
Thailand 60, 171
Thais, thaidids 79, 107
Thais kiosquiformis 78
Thalassia 141
Thalassina 44, 105
Thalassodendron 141
tides, effects of 3–4, 8–11, 18, 19, 31, 33–40,
 47–8, 89, 90, 98–9, 106, 113, 123–4, 136,
 141, 142, 147, 148, 160–1
tiger, Bengal (*Panthera tigris*) 73, 74
timber 170, 172, 175–6, 196, 197, 198
time allocation by crabs 90, 98–9, 101
trade-offs 16–18, 104
transpiration 17
tricarboxylic acid (TCA) cycle 10
Trichoderma 126
Trinidad 68, 71, 174
tsunami 83
turbellarians 111
typhoons, *see* hurricanes

Uca (fiddler crabs) 62, 84, 93–100, 102, 106,
 107, 114, 126, 129–30, 143, 149–50
 U. dussumieri 96–7, 99
 U. lactea 95, 96, 97, 98, 101, 107
 U. polita 95
 U. pugnax 97
 U. rosea 98, 99
 U. tangeri 74
 U. triangularis 99
 U. urvillei 98
 U. vocans 95
 feeding 86, 94–7

 reproduction 98
 social behaviour 93–4
Ucides 83, 92
Ulva 181
understorey vegetation 7, 46, 91, 176
United States 185
urea 63–4, 115
uses of mangroves, *see* exploitation

Vanda 51
Varanus indicus 66, 174
Vellar estuary 107
Verbenaceae 157
Vietnam 171, 179, 183
vivipary 24–32

waders 68
warblers 72
wasps 165
water conservation 102, 109, 115
water flow through burrows 50, 106
water uptake and use 12–14, 16–18
water use 186
waterlogging of soil 3–11, 18, 32
 see also anaerobic conditions
wave action 34, 35, 42
weaver ant 57, 61
West Indies 35
Wetherellia 158
whistler 70
white-eye 70
wood 46, 57, 127–8, 131
wood-borers 78–9, 109
woodpeckers 69

Xanthidae 92, 129
Xeromys myoides 74
Xiamen, China 119
xylem 52
Xylocarpus 2, 8, 13, 14, 24, 26, 27, 187
 X. australensis 53
 X. australoasicus 56
 X. granatum 6, 37, 53, 56
 X. mekongensis 6
 X. moluccensis 37

yellow fever 59

zinc 180
zonation 36–45, 78, 99, 188